SOFTWARE-IMPLEMENTED HARDWARE FAULT TOLERANCE

SOFTWARE-IMPLEMENTED HARDWARE FAULT TOLERANCE

O. Goloubeva, M. Rebaudengo, M. Sonza Reorda, and M. Violante

Politecnico di Torino – Dipartimento di Automatica e Informatica

 Springer

Olga Goloubeva, Maurizio Rebaudengo,
Matteo Sonza Reorda, and Massimo Violante

Politecnico di Torino
Dip. Automatica e Informatica
C.so Duca degli Abruzzi, 24
10129 Torino, ITALY

Software-Implemented Hardware Fault Tolerance

Library of Congress Control Number: 2006925117

ISBN-10: 0-387-26060-9   ISBN-10: 0-387-32937-4 (e-book)
ISBN-13: 9780387260600   ISBN-13: 9780387329376 (e-book)

Printed on acid-free paper.

Printed in the United States of America.

9 8 7 6 5 4 3 2 1

springer.com

Preface

Processor-based systems are today employed in many applications where misbehaviors can endanger the users or cause the loss of huge amount of money. While developing such a kind of safety- or mission-critical applications, designers are often required to comply with stringent cost requirements that make the task even harder than in the past.

Software-implemented hardware fault tolerance offers a viable solution to the problem of developing processor-based systems that balance costs with dependability requirements but since many different approaches are available, designers willing to adopt them may have difficulties in selecting the approach (or the approaches) that best fits with the design's requirements.

This book aims at providing designers and researchers with an overview of the available techniques, showing their advantages and underlining their disadvantages. We thus hope that the book will help designers in selecting the approach (or the approaches) suitable for their designs. Moreover, we hope that researchers working in the same field will be stimulated in solving the issues that still remain open.

We organized the book as follows. Chapter 1 gives the reader some background on the issues of fault and errors, their models, and their origin. It also introduces the notion of redundancy that will be exploited in all the following chapters.

Chapter 2 presents the approaches that, at time of writing, are available for hardening the data that a processor-based system elaborates. This chapter deals with all those errors that modify the results a program computes, but that do not modify the sequence in which instructions are executed.

Chapter 3 concentrates on the many approaches dealing with the problems of identifying the errors that may affect the execution flow of a program, thus changing the sequence in which the instructions are executed.

Chapter 4 illustrates the approaches that allow developing fault-tolerant systems, where errors are both detected and corrected.

Chapter 5 presents those approaches that mix software-based techniques with ad-hoc developed hardware modules to improve the dependability of processor-based systems.

Finally, chapter 6 presents an overview of those techniques that can be used to analyze processor-based systems to identify weakness, or to validate their dependability.

Authors are listed in alphabetic order.

Contents

Contributing Authors

Dr. Olga Goloubeva
Politecnico di Torino – Dipartimento di Automatica e Informatica
C.so Duca degli Abruzzi 24
10129 Torino, ITALY
E-mail: olga.golubeva@polito.it

Prof. Maurizio Rebaudengo
Politecnico di Torino – Dipartimento di Automatica e Informatica
C.so Duca degli Abruzzi 24
10129 Torino, ITALY
E-mail: maurizio.rebaudengo@polito.it

Prof. Matteo Sonza Reorda
Politecnico di Torino – Dipartimento di Automatica e Informatica
C.so Duca degli Abruzzi 24
10129 Torino, ITALY
E-mail: matteo.sonzareorda@polito.it

Dr. Massimo Violante
Politecnico di Torino – Dipartimento di Automatica e Informatica
C.so Duca degli Abruzzi 24
10129 Torino, ITALY
E-mail: massimo.violante@polito.it

Chapter 1

BACKGROUND

1. INTRODUCTION

Today we are living in a world where processor-based systems are everywhere. Sometimes it is easy to recognize the presence of a processor-based system, like in the automatic vending machine where we can select and buy the train ticket we need, the digital kiosks where we can post our digital pictures for printing, as well as in the desktop or laptop computer we have. Sometimes the presence of a processor-based system may not be so easily recognizable, like in the electro-mechanical unit controlling the operations of the engine or brakes of our car, the opening and closing of the car's windows, or even the microwave oven we use to warm the dinner.

Very often, processor-based systems have been already used to implement safety- or mission-critical applications, where any failure may have or already had dramatic impacts in terms of loss of money or of human lives. The recent history already recorded several cases where a problem within processor-based systems caused dramatic outcomes, as for example the computer-controlled Therac-25 machine for radiation therapy that caused the massive overdose of six patients [1].

A processor-based system can be used in safety-critical applications in several ways [2]. It may provide information to a human controller upon request. It may interpret data and display it to the controller, who makes decisions on them. It may issue commands directly, while a human operator controls the operations issued by the processor-based system, with the possibility of intervention on them. It may even replace the human control completely.

No matter the application scenario, the correct operations of the processor-based system are mandatory. In case a human operator has to take decisions on the basis of information produced by the processor-based system, he/she has to relay on the available information. More obvious is the case where the processor-based system is the only responsible for the operations carried out by the system.

As a result of the widespread adoption of processor-based system in mission- and safety-critical applications, there is an urgent need for developing products the user can reasonably rely on.

The literature makes available several approaches to meet such a demand, which are based on a massively use or redundant modules (redundant hardware, and/or redundant software) [3], and which have been developed to cope with the stringent dependability requirements of developers of traditional mission- or safety-critical applications: control systems for nuclear reactors, military and space applications.

Although effective in achieving the dependability requirements, these methods are becoming very difficult to be exploited today.

On the one hand, developers of traditional mission- or safety-critical applications have seen their budgets shrinking constantly, to the point that commercial-off-the-shelf components, which are not specifically designed, manufactured, and validated for being deployed in critical applications, are nowadays mandatory to cut costs.

On the other hand, the advancement in manufacturing technologies has set available deep-sub-micron circuits that pack tens of millions of transistors, operates in the GHz domain, and are powered by 1 Volt power supply, which open the frontiers for unprecedented low-cost computing power, but whose noise margins are so reduced that the obtained products are expected to experience (and some already are experiencing) problems when deployed in the field, even if they are designed, manufactured, and operated correctly. For such a kind of systems, which are intended for being deployed in commodity sea-level applications the correct behavior has to be guaranteed, although there are not necessarily mission- or safety-critical systems. For such a kind of commodity applications cost is one of the primary concerns, and therefore the techniques for guaranteeing high dependability coming from traditional critical domains are not affordable.

Finally, developers of mission- or safety-critical applications can benefit from the usage of commercial-off-the-shelf components not only for reasons related to component's cost, but also for performance reason. Indeed, commercial-off-the-shelf components are usually one generation behind their hardened counterparts (i.e., components that are certified for being deployed safely in critical applications), which means that they are in general more powerful, less power demanding, etc. As a result, developer of mission- or safety-critical application can produce better designs by using commercial-off-the-shelf components, provided that they have a low-cost way to design dependable systems from unreliable components.

The quest for reducing development costs, while meeting high dependability requirements, has seen the raise of a new design paradigm known as *software-implemented hardware fault tolerance* for developing processor-based systems the user can reasonably rely on. According to this paradigm, commercial-off-the-shelf processors are used in combination with specially crafted software. The processor executes the software whose purpose is twofold. It performs the original functionalities the designers implemented to satisfy the user's requirements, as well as monitoring functionalities that detect, signal, and possibly correct, the occurrence of hardware errors.

By using commercial-off-the-shelf processors, the designers can use the state-off-the-art components that guarantee the best performance available. Moreover, designers can cut costs significantly: commercial-off-the-shelf components come indeed at much lower costs than their hardened counterparts (even orders of magnitude lower). Moreover, hardware redundancy is not used, since all the tasks needed to provide dependability are demanded to the software.

According to this paradigm, the software becomes the most critical part of the system, since it is its duty to supervise the correct behavior of the whole system. On the one hand the software must be correct, i.e., it should implement correctly the specifications. Moreover, it should be effective in coping with errors affecting the underlying hardware.

The focus of this book is on describing the techniques available today for developing software that are effective in detecting, signaling, and (when possible) correcting hardware errors. The techniques needed for guaranteeing that the software is correct are out of the scope of this book and, although some of the techniques that will be presented can be also used for this purpose, they are not addressed here.

This chapter aims at giving the reader the background information needed for reading fruitfully the reminder of the book.

The chapter initially introduces some definitions. Then, it describes the error models that are used in the following of the book, as well as the main causes of transient faults in electronic systems. Finally, it introduces the readers with the concept of redundancy.

The intent of this chapter is to give to the reader an introduction to the issues that have stimulated the research community in the past years, and which are at the basis of the techniques we will describe in the following chapters.

2.   DEFINITIONS

2.1   Faults, errors and failures

Before exploring error models, it is important to introduce some terminology that will be exploited in the following of the book, and that is mainly taken from [2] and [4].

In this book we will refer to the scenario depicted in Fig. *1-1*, where we can recognize three components:

- The *system*, which is a physical entity that implements the functionalities needed by one or more *users*. In the following of this book the system will always be a processor-based systems, where software and hardware entities cooperate to implement the needed functionalities.
- The *user*, which is the entity that interacts with the system by providing input stimuli to the system, and by capturing and using the system's output responses. In this book, the user may be a human being, i.e., the user of a personal computer, as well as a physical entity, i.e., a processing module (the user) that reads and processes the data sampled by an acquisition module (the system).
- The *environment*, which is the set of external entities that may alter the system's behavior without acting on its inputs directly.

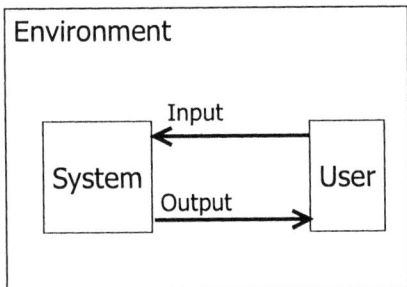

Figure 1-1. The scenario considered in this book, and its components: the systems, the user, and the environment

Due to the alterations induced by the environment (thermal stress, impact with ionizing radiations, etc...) a correct system may stop working correctly (either permanently, or by a period of time only).

In this book we will use the term *failure* to indicate the non-performance or inability of a system or a system's component to perform its intended function for a specified time under specified environmental conditions.

Given this definition, a *fault* can be defined as the cause of a failure. A fault can be *dormant* or *passive*, when it is present in a system, but the functioning inside the system is not disturbed, or it can be *active* when it has an effect on the system functioning. We refer to an active fault as an *error*. An error creates a failure as soon as it is propagated from the inside of the system to the system's outputs. As noted in [4], once a fault has been activated as an error in one system's module, several degradation mechanisms can propagate this error through the system's structure until the error reaches the system's outputs, thus producing a failure. This propagation process is conducted through *error propagation paths*, which depend on the system's module where the fault originates, the structure of the system, and the input sequence that is applied to the system.

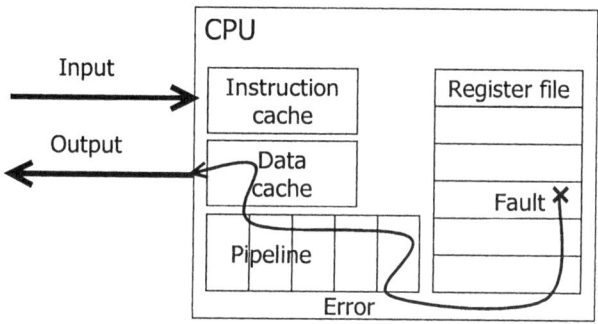

Figure 1-2. Example of error propagation

Fig. *1-2* reports an example of how a fault may propagate within a system up to the system's outputs. The considered system is a Central Processing Unit (CPU), whose Input is the program the CPU is executing, and its Output is the result of the executed instructions. In this example the fault is originated within one register in the Register file module. As a consequence of the program the CPU is executing, the fault remains passive for some clock cycles, after which it propagates through the Register file and it finally becomes active as an error when, as an example, it exits the Register file and enters the Pipeline module. After propagating through the different stages composing the CPU's Pipeline, the error affects the Data cache, and finally the CPU's Output, thus becoming a failure.

The same fault can produce different errors and different failures at different moments of the system's life. These effects depend on the fault location, and on the activity of the system during and after the fault's occurrence. In particular, faults do not necessarily become errors, and errors do not necessarily become failures. Fault propagation depends on the structure of the system and the input sequence that is applied to the system during its use. As an example, let us consider the code fragment reported in Fig. *1-3*.

```
if( R0 == R1 )
    result = f1( R0, R1);
else
    result = f2( R0, R1 );
```

Figure 1-3. A simple code fragment

Due to a fault, function `f1` is executed erroneously instead of `f2`. The following situations may be envisioned:
- The two functions `f1` and `f2` give the same result on the basis of the current value of `R0` and `R1`. In this case, the fault does not become an error, and thus the error propagation process is stopped.
- The two functions `f1` and `f2` give different results. In this case the fault becomes an error as soon as it propagates to the variable `result`, and it can possibly propagate through the system becoming a failure.

A fault may remain passive until an error is produced in a module of the system. The first occurrence of an error provoked by a fault is called *initial activation*. The term *latency* is used to identify the meantime between the fault occurrence and its initial activation as en error.

2.2   A taxonomy of faults

As suggested in [5], we can identify two types of faults: natural faults, and human-made faults.

Natural faults are faults affecting the hardware of a system that are caused by natural phenomena without human participation. In this type we may have *production defects*, which are natural faults that originate during development. During operation the natural faults are either *internal*, due to natural processes that cause physical deterioration, or *external*, due to natural processes that originate outside the system boundaries and cause physical interference by penetrating the hardware boundary of the system (e.g., radiation) or by entering via use interfaces (power transients, noisy input lines, etc.).

Human-made faults are the result of human actions, and may be *omission faults* when they are originated by absence of actions (that it is not performed when it should be), and *commission faults* when wrong actions are performed. If we consider the objective of the human interfering with the system, we can identify two further categories for the human-made faults:

- *Malicious faults*, which are introduced during either system development or during system use. The objective of who introduced malicious faults is to cause harm to the system during its use.
- *Nonmalicious faults*, which are introduced in the system without malicious intent. In this case we can find *nondeliberate* faults that are caused by mistakes of which the human interacting with the system (the developer, the operator, the maintainer) is not aware. We can also find *deliberate* faults that are caused by wrong intended actions.

2.3 Classifying the effects of faults

To describe the possible impact of faults in a processor-based system, we may refer to the following classification.

1. *Effect-less fault.* The fault does not propagate as an error neither as a failure. In this case the fault appeared in the system and remained passive for a certain amount of time, after which it was removed from the system. As an example, let us consider a fault that affects a variable x used by a program. If the first operation the program performs on x after x was affected by the fault is a write operation, then a correct value is overwritten over the faulty one, and thus the system returns in a fault-less state.
2. *Failure.* The fault was able to propagate within the system until it reached the system's output.
3. *Detected fault.* The fault produced an error that was identified and signaled to the system's user. In this case the user is informed that the task the system performs was corrupted by a fault, and the user can thus take the needed countermeasure to restore the correct system functionalities. In systems able to tolerate the presence of faults, the needed countermeasures may be activated automatically. Error detection is performed by means of mechanisms (*error-detection mechanisms*) embedded in the system whose purpose is to monitor the behavior of the system, and to report anomalous situations. When considering a processor-based system, error-detection mechanisms can be found in the processor, or more in general in the hardware components forming the system, as well as in the software it executes. The former are usually known as *hardware-detection* mechanisms, while the latter are known as *software-detection* mechanisms. As an example of the hardware-detection mechanisms we can consider the *illegal instruction trap* that is normally executed when a processor

decodes an unknown binary string coming from the code memory. The unknown binary string may be the result of a fault that modified a valid instruction into an invalid one. As an example of the software-detection mechanisms we can consider a code fragment the designers inserted in a program to perform a range check, which is used to validate the data entered by the systems' user, and to report an alert in case the entered data is out of the expected range. To further refine our analysis, it is possible to identify three types of fault detections:

- *Software-detected fault.* A software component identified the presence of an error/failure and signaled it to the user. As an example, we can consider a subprogram that verifies the validity of a result produced by another subprogram stored in a variable x on the basis of range checks. If the value of x is outside the expected range, the controlling subprogram raises an exception.

- *Hardware-detected fault.* A hardware component identified the presence of an error/failure and signaled it to the user. As an example, we can consider a parity checker that equips the memory elements of a processor. In case a fault changed the content of the memory elements, the checker identifies a parity violation and it raises an exception.

- *Time-out detected fault.* The fault forced the processor-based system in an unexpected state from which the system does not provide any output results (examples of this state are an endless loop, or the halt state processors usually have). This fault type can be detected by monitoring the processor's activities, and by signaling the occurrence idle periods longer than usual. As an example, the occurrence of this fault type may be detected thanks to a watchdog timer that is started at the beginning of the operations of the processor-based system, and that expires before the system could produce any result.

4. *Latent fault.* The fault either remained passive in the system, or it became active as an error, but it has not been able to reach the system's outputs, and thus it has not been able to provoke any failure. As an example, we can consider a fault that modifies a variable x after the last usage of the variable. In this case, x holds a faulty value, but since the program no longer uses x, the fault is unable to become active and propagate through the system. The fault/error may transform into a failure (or any other category) later in the system's lifetime.

5. *Corrected fault.* The fault produced an error that the system was able to identify and to correct without the intervention of the user. Corrected faults are indistinguishable from effect-less ones unless the

system collects and transmits to its user suitable status information informing that a fault was detected and later corrected.

At the end of the propagation process, if the error propagation has not been stopped, the fault transforms into a failure. As a result, the system does not deliver the expected functionality. As noted in [4], it is possible to identify four grades of consequences of failures:

- *Benign*. The failure has no serious consequences on the task the system performs.
- *Significant*. The task the system performs is disturbed and the efficiency of the delivered service is reduced.
- *Serious*. The task the system performs is disturbed greatly.
- *Catastrophic*. The task the system performs is stopped with the destruction of the controlled process, or with human injuries or deaths.

2.4   Dependability and its attributes

The term that is normally used to characterize a processor-based system involved in safety- or mission-critical applications is *dependability*, for which we can give two definitions [5]. The term dependability is defined as that ability of a processor-based system to deliver a service that can justifiably be trusted. Although effective, this definition mandates the capability of justifying the trust in a system, and thus it is somewhat subjective: one user may accept a delay of 1 second for providing a correct answer from a system intended for responding in 100 milliseconds, while another user may not accept this case. In this example the first user sees the system as dependable, while the system is not dependable for the second user. A more objective definition, which is presented in [5], states that a system is dependable if it is able to avoid service failures that are more frequent and more severe that is acceptable. Under this definition, being the system of the previous example able to provide a correct answer, we can consider it as dependable.

The dependability is a concept that integrates several attributes of a system:

- *Availability*, which is the readiness for a correct service. It can be also defined as the probability that a system is able to deliver correctly its service at any given time.
- *Reliability*, which the capability of providing the continuity of a correct service. It can be also defined as the probability of a system to function correctly over a given period of time under a given set of operating conditions.
- *Safety*, which is the capability of avoiding catastrophic consequences on the users or the environment.

- *Integrity*, which is the capability of avoiding improper alterations. As suggested in [6], we can define two types of integrity:
  1. *System integrity* defined as the ability of a system to detect faults in its own operations and to inform a human operator.
  2. *Data integrity* defined as the ability of a system to prevent damage to its own database and to detect, and possibly correct, errors that do occur as consequence of faults.
- *Maintainability*, which is the capability of undergo modifications and repairs. Alternatively, we can define the term *maintenance* as the action taken to retain a system in, or return a system to, its designer operating condition, and the *maintainability* as the ability of a system to be maintained.

3.   ERROR MODELS FOR HARDWARE AND SOFTWARE COMPONENTS

In this book we mainly consider faults and errors affecting the hardware of a processor-based system, only. We will thus present several techniques able to detect the occurrence of hardware faults, and when possible correct them, before they become failures. Although our discussions are focused on hardware faults/errors only, some of the techniques this book presents could be used effectively to deal with software errors (e.g., bugs), too. For this reason we will briefly outline in this section the most important error models introduced to account for software errors.

Given the behavioral properties of a system, i.e., the knowledge about the function the system performs, it is possible to define an *error model* as a set of faults that are active and originate an error.

3.1   Error models for hardware components

In the literature several error models can be found. Some of them are listed here, organized in several categories.

Error models can be obtained by observing the modifications faults introduce in the values the system manipulates. In this case we have:
- *Logical errors*: they are characterized by transformations of logical values. For example a '0' becomes a '1', or vice versa.
- *Non-logical errors*: they are characterized by transformations of logical values outside the specification domain. For example, the altered value is between '0' and '1'.

Other error models can be obtained by observing the time a fault needs to alter a fault-free system, thus having:

- *Static errors*: they correspond to stable undesired situations. For example the output of a gate is '1' instead of '0'.
- *Dynamic errors*: they correspond to transient and unstable undesired situations. For example the output of a gate oscillates before reaching a correct and stable value.

Moreover, other error models may be defined by observing the duration of faults, thus having:

- *Hard errors*: they correspond to permanent modifications to the expected functioning of systems. For example, the output of a gate is stuck at '0' or '1'.
- *Soft errors*: they correspond to temporary modifications to the expected functioning of systems. For example, a memory element stores a wrong '1' value instead of correct '0' value for one clock cycle.

Finally, we can define error-models by observing the multiplicity of effects produced by faults in a system, thus having:

- *Single errors*: they disturb only one element of a system.
- *Multiple errors*: they disturb several elements of a system.

In order to define the hardware error models used in this book, we adopted a two-tier hierarchical approach. At the bottom of the hierarchy lie the hardware components implementing the processor-based system (i.e., memory modules, arithmetic unit, control unit, etc.). At the top of the hierarchy we find the information the system handles: program's data, and program's instructions. The elements of the two hierarchy levels we considered are outlined in Fig. *1-4*.

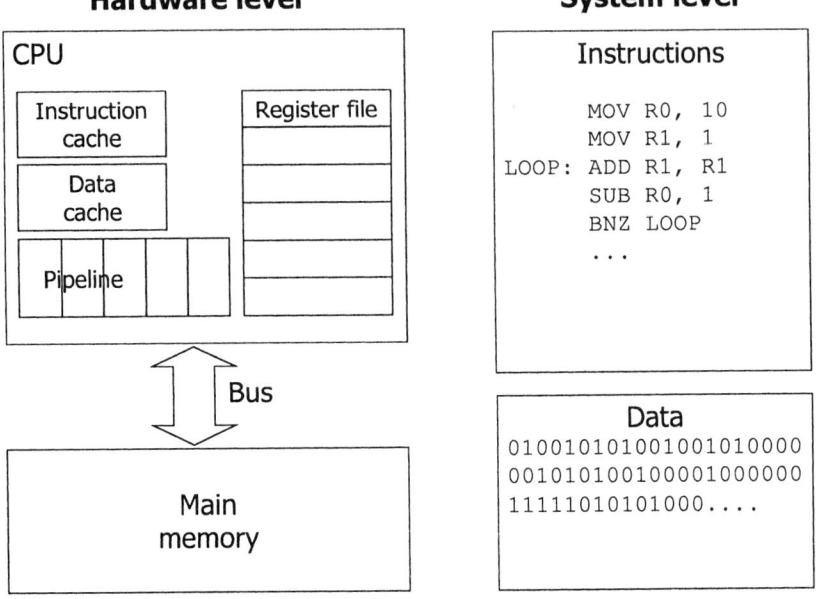

Figure 1-4. The elements in the two levels of our abstraction hierarchy

As far as the hardware level is considered, we consider the systems as composed of three main elements: the central processing unit (CPU), the main memory, and the bus connecting them. In order to analyze fault effects more carefully, we further partition the CPU in its main components. We adopted as a reference model that of a modern Reduced Instruction Set CPU (RISC CPU), where the main components are: the instruction cache, the data cache, the pipeline and the register file. Please note that the considerations reported in the following of this chapter, although based on this reference model, can be extended easily to different CPU's architectures, spanning from simpler ones (like those of not-pipelined microcontrollers) to more complex ones (like those of superscalar processors).

As far as the system level is considered, its two components are the data the program manipulates, and the program's code.

3.1.1   Hardware-level error models

When considering the system's hardware we can define hardware-level error models, no matter which type of function the hardware implements. In this book we consider the following hardware-level error models that can be described as a combination of the previously introduced ones:

- *Single stuck-at*: it is defined as logical, hard and single error resulting from hardware faults that affect system's components. As an example, let us consider the system component *C* depicted in Fig. *1-5* having one out *Cout* and one input *Cin*. In case a single stuck-at affects the component, *C* may have *Cout* permanently stuck either at 1 or at 0, and the same may happen to its input *Cin*.
- *Single bit-flip*: it is defined as logical, soft and single error resulting from hardware faults that alter one of the system's memory elements. When the memory element is affected by the bit-flip its content is changed from 1 to 0 or vice-versa.

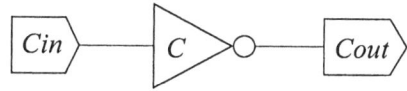

Figure 1-5. A simple component of the system

Hardware faults leading to single stuck-at errors are well known, and discussions about them can be found in [7]. Conversely, hardware faults causing single bit-flips, known as single-event effects, are relatively new, and they are becoming of great interest today due to the evolution of semiconductor manufacturing technologies [8]. For these reasons, we leave the discussion of hardware faults causing stuck-at to other texts (like [7]), and we will address in the sections 4, and 4.3 single-event effects, only.

3.1.2   System-level error models

When considering the information the system handles we can identify the following system-level error models:

- *Single data error*: it is defined as a single logical error affecting the program's data stored in the system. Please note that this definition does not consider the location in the system where the data are actually stored: they may be stored either in the system's main memory, or in the processor's data cache, or in the processor register file.
- *Single code error*: it is defined as a single logical error affecting one instruction of the program's code. As previously done, we do not consider where the erroneous instruction is located within the system: it

may either be in the system's main memory, or in the processor's instruction cache, or in the processor's pipeline. In order to model accurately the errors that may affect one instruction of the program's code, we defined two types of code errors:

```
%% Error-free code

        MOV R0, 10
        MOV R1, 1
LOOP:   ADD R1, R1
        SUB R0, 1
        BNZ LOOP
```

```
%% Erroneous code

        MOV R0, 10
        MOV R1, 1
LOOP:   SUB R1, R1
        SUB R0, 1
        BNZ LOOP
```

Figure 1-6. Example of system-level code error of type 1, where an ADD instruction is modified in a SUB instruction. The hardware-level error modified the code of the original instruction in such a way that it was transformed in another-one, but no change to the program flow is introduced.

- *Type 1*: it is defined as a single code error that modifies the operation the instruction executes, but that does not change the expected program's execution flow. Examples of this error model are reported in Fig. *1-6* and Fig. *1-7*: a first example is given where an ADD instruction is replaced with a SUB one, and a second example is given where the addressing mode of the instruction is changed from an immediate addressing to a direct addressing. We remark that in both the examples, the expected program's execution flow is not modified.

```
%% Error-free code

        MOV R0, 10
        MOV R1, 1
LOOP:   ADD R1, R1
        SUB R0, 1
        BNZ LOOP
```

```
%% Erroneous code

        MOV R0, 10
        MOV R1, 1
LOOP:   ADD R1, [R1]
        SUB R0, 1
        BNZ LOOP
```

Figure 1-7. Example of system-level code error of type 1, where the addressing mode of an ADD instruction is modified. The hardware-level error modified the code of the original instruction in such a way that it was transformed in another-one, but no change to the program flow is introduced.

- *Type 2*: it is defined as a single code error that modifies the expected program's execution flow. Examples of this error models are reported in Fig. *1-8* and *1-9*: a first example is given where the displacement field of a branch instruction is changed, and a second example is given where the condition upon which a conditional branch is taken is changed.

```
%% Error-free code

          MOV R0, 10
          MOV R1, 1
LOOP: ADD R1, R1
          SUB R0, 1
          BNZ LOOP
```

```
%% Erroneous code

          MOV R0, 10
          MOV R1, 1
LOOP: ADD R1, R1
          SUB R0, 1
          BNZ elsewhere
```

Figure 1-8. Example of system-level code error of type 2, where the target address of a branch is changed. In this case, the hardware-level error modified the code of the original instruction in such a way that the expected program's execution flow is changed.

```
%% Error-free code

          MOV R0, 10
          MOV R1, 1
LOOP: ADD R1, R1
          SUB R0, 1
          BNZ LOOP
```

```
%% Erroneous code

          MOV R0, 10
          MOV R1, 1
LOOP: ADD R1, R1
          SUB R0, 1
          BZ LOOP
```

Figure 1-9. Example of system-level code error of type 2, where the branch condition of a conditional branch is changed. Again, the hardware-level error modified the code of the original instruction in such a way that the expected program's execution flow is changed.

3.1.3   Hardware-level errors vs. system-level errors

Although the software techniques to harden processor-based systems presented in the following of this book aims at detecting, and when possible correcting, hardware-level errors, they have been developed by researchers reasoning on system-level errors. System-level errors are an abstraction of hardware-level errors that simplify the task of researchers that can work to a level closer to that of programs and programs' data, while still providing a good modeling accuracy: in most cases system-level errors correspond indeed to hardware-level ones.

As an example, let us consider a hardware-level single bit-flip affecting the memory elements of a processor-based system. The following situations can be envisioned, depending on the affected components.

- *System's main memory.* The main memory of a processor-based system is normally organized in at least three segments: the data segment (storing program's data), the code segment (storing program's instructions), and the stack segment (storing the program's stack). According to the segment the single bit-flip affects, we can identify the corresponding system-level errors. The possibilities are:

- *Data segment*: the hardware-level single bit-flip error in the memory area storing the data segment corresponds to a system-level single data error.

- *Code segment*: the hardware-level single bit-flip error in the memory area storing code segment corresponds to a system-level single code error. It may be either of type 1 or of type 2 depending on the location of the bit-flip in the instruction. As an example of a hardware-level single bit-flip producing a system-level code error of type 1, let us consider Fig. *1-10*. In Fig. *1-10* the format of a SPARC v9 data-manipulation instruction is reported, where the field named rd is a binary code that specifies which register in the register file is affected by the instruction. Any hardware-level single bit-flip changing the rd value produces a system-level single code error of type 1, since it changes the register the instruction affects, without changing the expected program's execution flow.

To illustrate an example of a hardware-level single bit-flip corresponding to a system-level code error of type 2, let us consider Fig. *1-11*, where the format of a SPARC v9 branch instruction is reported. The field named disp30 is a binary code specifying the relative address where the program execution should continue. Any hardware-level single bit-flip in this field will modify the expected program's execution flow, thus corresponding to a system-level code error of type 2.

31	29	24	21		0
op	rd	op2		imm22	

Figure 1-10. The format of a data-manipulation instruction according to the SPARC v9 instruction set

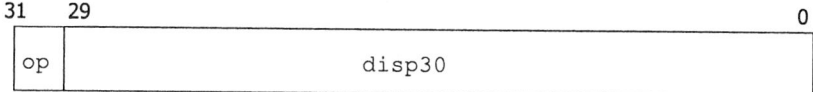

31	29		0
op	disp30		

Figure 1-11. The format of a branch instruction according to the SPARC v9 instruction set

- *Stack segment*: the hardware-level single bit-flip error in the memory area storing stack segment corresponds either to a system-level data error (in case the affected memory location corresponds to data exchanged between procedure calls or temporary variables), or to a system-level code error of type 2 (in case the affected memory location corresponds to an address stored in the stack by a procedure call).
- *Processor's data cache.* Similarly to what happened for the memory's data segment, any hardware-level single bit-flip error in the data cache corresponds to a system-level data error.
- *Processor's instruction cache.* Similarly to what happened for the memory's code segment, any hardware-level single bit-flip error in the instruction cache corresponds to a system-level code error either of type 1 or of type 2.
- *Processor's register file.* Hardware-level single bit-flip errors may correspond to the following types of system-level errors, depending on the affected register, as well as the processor's architecture:
  - Single data error, in case the bit-flip affects a register storing the data the program elaborates.
  - Single code error of type 1, in case the register contains an address used by a load/store instruction.
  - Single code error of type 2, in case the register contains the address of a branch target.
- *Processor's control registers.* Hardware-level single bit-flip errors affecting processor's control registers may correspond to any type of system-level errors. We may have:
  - Single data error in case the bit-flip modifies a temporary register used by computations, or a forwarding register used by a pipelined computing unit. As an example, let us refer to Fig. *1-12*, which reports the conceptual architecture of a RISC CPU, showing the layout of the CPU pipeline's five stages. In this architecture, registers named Op1, Op2, m.result, and w.result store the operands and the results needed and produced by data-manipulation instructions. Any hardware-level single-bit flip in these registers corresponds to a system-level single data error since it affects program's data.

- Single code error of type 1 in case the bit flip modified the processor's
  instruction register, or a boundary register within the pipeline storing
  the instruction code. Examples of this correspondence can be found in
  Fig. *1-12* by considering the hardware-level single bit-flip in the
  registers named d.instr, e.instr, m.instr, and w.instr which
  contain the binary code of the instructions in the CPU's pipeline.
  Some of these hardware-level errors (i.e., all those bit-flips that do not
  transform a data-manipulation instruction in a branch one, or a branch
  one in a different one) correspond to system-level code errors of type 1
  since they affect, and possibly change, the instructions that the CPU
  executes.

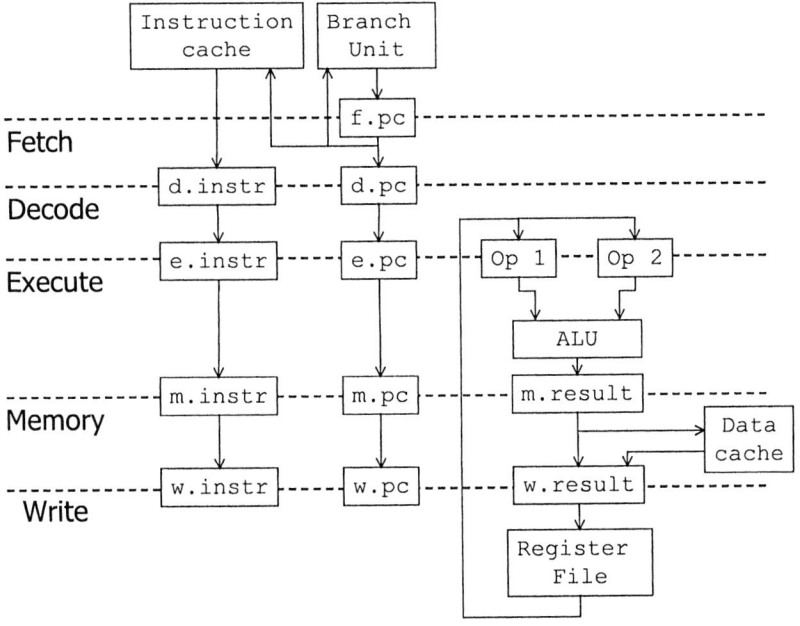

Figure 1-12. Conceptual architecture of a RISC CPU

- Single code error of type 2 in case the bit-flip modifies the instruction
  register, or a boundary register of the pipeline, changing the target of a
  branch, or the condition of a conditional branch. The same type or error
  may be produced by bit-flip in the processor's status word, or in the
  program counter. Example of this type of hardware-level errors can be
  found in Fig. *1-12* by considering the registers named d.instr,

`e.instr`, `m.instr`, and `w.instr`, and those named `f.pc`, `d.pc`, `e.pc`, `m.pc`, and `w.pc`. The former may be subject to hardware-level single bit-flips that change data-manipulation instructions in branch ones, or that change the register's bits defining upon which condition the branch should be taken or not. The latter may be subject to hardware-level single bit-flips that modify the processor's program counter. In both cases, these hardware-level errors correspond to system-level code errors of type 2.

When needed, in the following of this book we will discuss the capabilities of software detection and correction techniques in terms of system-level errors, and when required we also provide hints on the corresponding hardware-level errors.

3.2   Error models for software components

Although this book addresses specifically hardware faults, some of the techniques that will be presented in the following chapters are useful to cope with software faults, too. For this reason we present here a brief discussion about error models for software components.

The general error models introduced in section 3.1 for hardware components are applicable to software components, too; in particular, the following parameters are relevant:

- *Static* or *dynamic errors*. In order to describe this type or error, let us refer to an example taken from [4]. Let us consider a system which handles sampled data coming from a sensor, which acquires a new input value every 10 ms and stores it in a variable called x. If the program that implements the sampling function sets x to a *null* value at the end of the usage of the last sampled value, the variable x will store an incorrect value until a new sample is acquired. This is an example of dynamic error since the variable x stores an incorrect value only for a limited amount of time, after which its value becomes correct as a new sample is acquired. An example of a static error would be an erroneous analog to digital conversion that would store a wrong value in x.

- *Permanent* or *temporary errors*. Let us consider a multi-tasking environment, where a task T1 uses a shared variable x written by a second task T2 [4]. When the program execution starts, x is not assigned to a correct value. If T2 assigns to x a correct value before T1 reads it, no problems occur. Conversely, if T1 reads the value before is has been assigned by T2, an error occurs. If the tasks T1 and T2 are cyclically executed, the error disappears at the next cycle (after x has been assigned by T2). Conversely, if x is read by T1 only once, the error is permanent.

- *Single* or *multiple errors*. Single errors are errors that affect only one element of the software component [4]. The term element depends on the model used to describe the component. At the programming level, it may be a variable, or a function. At the system level, it may be an object, or a resource. Conversely, multiple errors occur when several elements are affected.

3.2.1   Error models at the source-code level

A program is a structure made up of an assembly of features provided by a language. These features are defined by their *syntax*, allowing fault models to be expressed, and their *semantics* specifying their behavior.

The negation of the properties associated with the semantics of a programming language defines an error model: negating the language's semantic is indeed general and it is applicable to any program independently from its functionality. The following five examples, taken from [4], can be used to clarify this issue.

- Let us consider a programming language that defines functions, and well as procedures. Both are subprograms, but the former is expected to return a value, accordingly with the language's semantic, while the latter is not expected to return a value. In this case, a possible error is a function that does not return a value. Several faults may be at the origin of this error. It may be a human-made omission fault due to the negligence of the programmer: the programmer forgot to write the return statement. It may also be an external natural fault: the return statement may exist, but a fault resulted in a control flow path that does not conclude the execution of the function with the execution of the expected return statement.
- An input parameter of a subprogram is not assigned by an actual value at subprogram call. As an example, this error occurs if a call push (X) is called with a non-initialized value of X. Similarly, this error may be the result of either a human-made fault or a natural fault.
- An output parameter of a subprogram is not assigned at the subprogram body execution completion. For instance, no value is returned in Y after the execution of pop (&Y). As before, this may be due to the negligence of the programmer, or due to the environment.
- A variable whose type is constrained is assigned by a value not belonging to the range specified by this type. In this case, the error is likely to be caused by the programmer.
- A first task calls the service of a second task that does not exist. This occurs when the second task was not previously created or if, when

being created, it was then terminated. The potential faults that are at the origin of this last error are many:

- The source program design explicitly express that the second task must be completed before the call.
- The second task was terminated due to an error raised during its execution.
- The second task was unintentionally terminated by another task.

Some researchers analyzed the properties of a given programming language, and identified a set of error models at the source-code level. As an example, [9] reports a study of the C language, and proposes the following error models that the programmer may introduce into a program:

- *Errors affecting assignments*: it is provoked either by missing or wrong local variable assignments.
- *Errors affecting conditional instructions*: it is provoked by one of the following faults:
  - An assignment statement is coded instead of a comparison one (e.g., if(a=b) instead of if(a==b)).
  - A wrong Boolean function is coded (e.g., if(a<b) instead of if(a<=b)).
  - A wrong number of iteration is coded (e.g., while(a<b) instead of while(a<=b))).
- *Errors affecting function call/return*: it is provoked by one of the following faults:
  - Coding of the wrong usage of parameters in function calls.
  - Omission of the needed return statement.
  - Coding of the wrong return statement.
- *Errors affecting algorithms*: it is provoked by one of the following faults:
  - Miss aligned else due to erroneous use of parenthesis.
  - Usage of binary operators instead of logical ones (e.g., a & b instead of a && b).
  - Coding of wrong Boolean expressions due to erroneous use of operator's precedence.
  - Missing statements.
  - Missing function calls.

3.2.2   Error models at the executable-code level

As before, the error models highlight the violations of expected properties, which now concern the executable code. As example of such a kind of error models, we may refer to the case where the execution of a subprogram is not terminated by a return instruction. This instruction is mandatory to restore the caller context. Several causes can be at the origin of this error. For instance, it may be due to the execution of a jump instruction of the subprogram body whose associated address was corrupted. The fault that provoked such a situation may be:

- A bad expression used to calculate the branching address due to a compiler failure.
- A bad constant address coming from an erroneous memory word where this data is stored.

The execution stack overflow is a second example of this error model. A stack is used at runtime to manage subprogram calls, to handle interruptions, etc. Various faults can be at the origin of this class of errors:

- Infinite recursion of a subprogram due to bad design or programming.
- Bad assessment of the stack memory size due to the compiler whose generated code does not optimize the stack use, or the runtime execution environment (e.g., the operating system) that does not master correctly the dynamic memory allocation.

4.   ORIGIN OF SINGLE-EVENT EFFECTS

Single-event effects arise when highly energized particles present in natural space environment strike sensitive regions of circuits. Depending on several factors, the particle-strike may cause no observable effect, a transient disruption of circuit's operation, a change of logic state, or a permanent damage to the integrated circuit [10].

In Sub-section 4.1, we will describe the source of highly energized particles, while in Sub-section 4.2 we will describe the physical origin of single-event effects.

4.1   Sources of highly energized particles

The sources of highly energized particles can be classified in different ways, depending on where the system is deployed. We can consider three so-called *radiation environments*: space, atmospheric, and ground radiation environments [11].

4.1.1   Space Radiation Environment

The space radiation environment is composed of two types of particles: particles trapped by planetary magnetospheres in "belts", which include protons, electrons, and heavier ions, and transient particles that include protons and heavy ions of all the elements of the periodic table. The transient particles belong to transient radiations, which consist of galactic cosmic ray particles and particles from solar events, such as coronal mass ejection and flares. These two types of solar eruptions produce energetic protons, alpha particles, heavy ions, and electrons [11].

Table *1-1* reports the maximum energies of particles that can be observed in the space radiation environment. Energies are expressed by using the eV (electron volt unit of measure). By definition, a single electron that is accelerated though a potential differential of one volt gains a kinetic energy of 1 eV, which is equivalent to $16 \cdot 10^{-21}$ joules.

Table 1-1. Maximum Energies of Particles

Particle Type	Maximum Energy
Trapped Electrons	10 MeV
Trapped Protons and Heavy Ions	100 MeV
Solar Protons	1 GeV
Solar Heavy Ions	1 GeV
Galactic Cosmic Rays	1 TeV

As remarked by the authors of [11], the space radiation environment is composed of particles with very high energy, and therefore shielding may not be effective in protecting circuits.

4.1.2 Atmospheric radiation environment

When cosmic ray and solar particles enter the Earth's atmosphere, they are attenuated by interactions with atoms of nitrogen and oxygen. The attenuation process produces protons, electrons, neutrons, heavy ions, muons, and pions. Among them, the most important ones are neutrons, which are present in measurable quantities starting from 330 Km of altitude. Their density increases with decreasing altitude, and it reaches its peak density at about 20 Km of altitude. Below than 20 Km, the neutron density starts to decrease, and at the ground level its density is about 1/500 of the peak one [12].

The maximum energy observed for the particles in the atmospheric radiation environment is about some hundreds of MeV.

4.1.3 Ground radiation environment

At the ground level both natural and man-produced radiations are present. Beside nuclear facilities, the most important source of radiations are galactic cosmic rays, which are capable of inducing single event effects.

Cosmic radiation at the ground level is the product of several generations of interactions of galactic cosmic rays and solar particles in the atmosphere. The density of radiations is strictly related with the 11-year solar cycle that modulates the density of galactic cosmic rays, and it can increases up to 5000% during large solar events.

4.2   Physical origin of single-event effects

Radiations can interact with materials producing two types of interactions: atomic displacement and ionization. The former corresponds to modifications to the structure of struck materials, which may show for example displaced atoms, and it is out of the scope of this chapter. Conversely, the latter corresponds to the deposition of energy in the struck materials [13], and it is focused in this chapter.

Ionizing radiations may interact with a circuit through two methods: direct ionization by the particle that strikes the circuit, or ionization by secondary particles created by nuclear reactions between the incident particle and the struck circuit. Both methods are critical, since both of them may produce malfunctions to the struck circuit [10].

4.2.1   Direct ionization

When an energetic particle passes through a semiconductor material it frees electron-hole pairs along its path, and it loses energy. When all its energy is lost, the particle rests in the semiconductor, after having traveled a path length that is known as *particle's range*. The term that is often used to describe the energy loss per unit path length of a particle as it passes through a material is *linear energy transfer* (LET). The unit of measure of LET is MeV/cm^2/mg: the energy loss per unit path length (whose unit of measure is MeV/cm) is indeed normalized by the density of the traversed material (whose unit of measure is mg/cm^3) so to be able to express the LET independently by the traversed material. The LET of a particle can be related quite easily to the charge it deposits into the traversed materiale. In silicon, an LET of 97 MeV/cm^2/mg deposits a charge of 1 pC/μm.

Direct ionization is the primary charge deposition mechanism for upsets caused by heavy ions (i.e., any ion with atomic number grater then or equal to two) [10]. Lighter particles such as protons do not usually produce enough charge by direct ionization to cause single-event effects. However, recent studies showed that as devices become smaller and thus more sensitive to particles, single-event effects due to direct ionization by means of protons are possible [14][15].

4.2.2 Indirect ionization

Indirect ionization is the primary mechanism through which light particles, such as protons and neutrons, may produce single-event effects. As a high-energy proton, or a neutron, enters a semiconductor lattice it may have an inelastic collision with atom's nucleus, provoking one of the following nuclear reactions: elastic collision that produces silicon recoils, the emission of alpha or gamma particles and the recoil of a daughter nucleus, and spallation reactions, in which the target nucleus is broken into two fragments, each of which can recoil. Any of these reaction products can deposit energy along their paths by direct ionization. Because these particles are much heavier than the original proton or neutron, they deposit higher charge densities as they travel the semiconductor, and therefore they may be capable of causing single-event effects [10].

4.3 Single-event effects in memory circuits

Single-event effects in memory circuits have the macroscopic effect of changing the content of a memory bit, provoking the so-called *Single Event Upset* (SEU). When ionizing radiations hit a memory circuit, the injected charge may indeed change the status of one bit that flips either from 1 to 0, or vice versa.

The SEU generation mechanisms are different depending on the memory's technology. Section 4.4 presents how SEUs may be generated within dynamic random access memories (DRAMs), which usually are the building blocks of the main memory in processor-based systems. Conversely, section 4.5 presents the generation mechanisms for static random access memories (SRAMs), that are the building blocks for the memory elements processors embed for implementing instruction and data caches, register file, and internal registers (control registers, pipeline boundary registers, etc.).

4.4 SEU mechanisms in DRAMs

As explained in [10], DRAM technology refers to the broad class of information storage devices, usually one-transistor designs, which store passively packets of charge to represent binary information. The key to understand the SEU generation mechanisms in DRAMs is that the information storage is passive (indeed no active information regeneration path exists), and any disturbance of any magnitude of the stored information provoked by ionizing radiations is persistent until it is corrected by a new write operation. In DRAMs there is no inherent refreshing of this charge packet, and no active regenerative feedback exists. As a result, a degeneration of the stored charge packet corresponding to a signal level outside the noise margin of the read circuit is sufficient to lead to erroneous interpretation of the stored information.

Two parameters are related to DRAM SEUs: the noise margin associated with a bit signal and a critical time window (since DRAM is a dynamic circuit, its sensitivity to SEUs changes with time). The noise margin is related with the concept of critical charge, Q_{crit}. Q_{crit} is usually defined as the minimum amount of charge collected at a sensitive node that is necessary to cause a circuit to change its state (i.e., to upset).

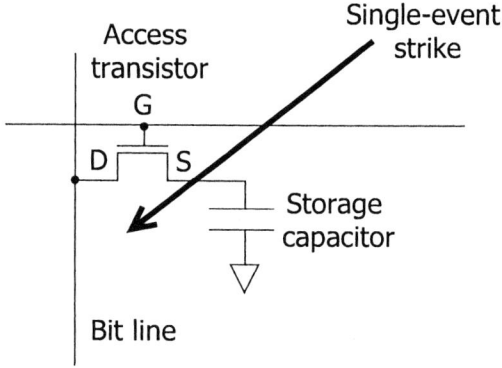

Figure 1-13. A cell of a DRAM array and its SEU generation mechanism.

The most prevalent SEU source in DRAMs is the single-event charge collection within each binary cell forming the DRAM array. These cell errors are caused by a single-event strike in or near either the storage capacitor or the source of the access transistor, as shown in Fig. *1-13*. Such a strike affects directly the stored charge and the information integrity by the collection of induced charge. A cell upset due to charge collection is usually observed as a 1 to 0 transition [16]. A further effect known as ALPEN [17] was later observed, which consists in the shunting of charge onto the storage capacitor. Thus a 0 to 1 transition can also be introduced by single-event strike.

SEUs can also occur in DRAMs due to bit-line strikes. When bit-lines are in a floating voltage state (e.g., due a read cycle), DRAMs are sensitive to the collection of charge into diffusion regions that are electrically connected to the bit line. This collection could arise from any of the access-transistor drains along the bit-line length or from a direct strike to the differential sense amplifier. The bit-line SEU mechanism is the reduction of the sensing signal due to a charge imbalance introduced on the precharged bit lines, either prior to or during the sensing operation [10][18].

Bit-line strikes are only possible during the floating precharge and sensing stages of operation, and therefore temporal characteristics of the strike in relation to the clocking of the DRAMs are critical. Because the duty cycle of these stages to the overall cycle time increases with increasing the overall clock frequency, the bit-line soft error rate is inversely proportion to DRAM cycle time. Conversely, cell upsets are independent of the DRAM cycle time. Bit-line errors also show a strong inverse correlation with the signal charge. As chip densities and speeds grow, bit-line errors are expected to be increasingly important [10].

A different failure mode was observed in 1988, due to a synergetic effect of bit-line and storage cell charge collection [19]. Both processes individually resulted in less charge collection than Q_{crit}, but the combined effect during a read operation caused an error. This effect, called *combined cell-bit line* (CCB) failure mode, was shown to dominated both the cell and bit-line error components at very low cycle time.

Another very important factor in determining the SEU sensitivity of DRAMs is the storage cell technology [20].

4.5   SEU mechanisms in SRAMs

The SEU generation mechanisms in SRAMs is quite different from DRAMs, due to the active feedback in the cross coupled inverter pair that forms a typical SRAM memory cell, as shown in Fig. *1-14*. When ionizing radiations strike a sensitive location in a SRAM (typically the reverse-biased drain junction of a transistor biased in the "off" state, the "off" n-channel transistor in Fig. *1-14*), charge collected by the junction results in a transient current in the struck transistor. As this current flows through the struck transistor, the restoring transistor ("on" p-channel transistors in Fig. *1-14*) sources current in an attempt to balance the radiation-induced current. The restoring transistor has a finite amount of current drive, and a finite channel conductance. Current flow through the restoring transistor therefore induces a voltage drop at its drain. This voltage transient in response to the single-event current transient is actually the mechanisms that can cause SEU in SRAM cells. The voltage transient is similar to a write pulse and can cause the wrong memory state to be latched into the memory cell.

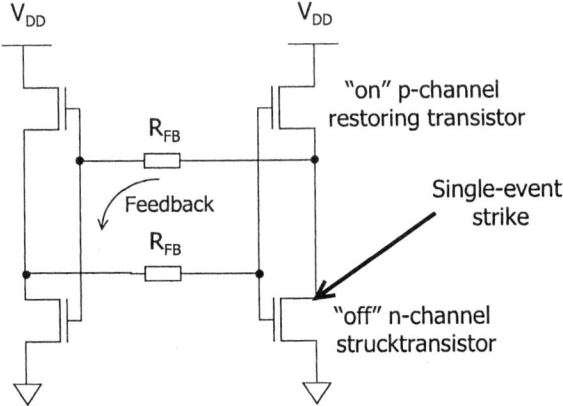

Figure 1-14. A cell of a SRAM array and its SEU generation mechanism.

SRAM cells have four possible sensitive strike locations corresponding to the four transistors' drains interior to the SRAM circuit [10].

4.6   Single-event effects in logic circuits

Due to the dramatic shrinking of devices' feature size, the reduction of power supply, as well as the increase of operating frequency, the noise margin of today logic circuits is extremely reduced. Although these technology advancements can be beneficial from the performance point of view (more transistor can be fit in a die, allowing systems performing more functions, quicker, and with less power consumption) they can have dramatic drawbacks from the dependability point of view.

In a logic circuit charge collection due to a single-event strike may generate a low-to-high or high-to-low voltage transition on a circuit line. This transition is known as *Single Event Transient* (SET), and it may provoke circuit misbehaviors in case its magnitude is compatible with the circuit's voltage swing.

A SET is originated when highly energized particles strike a sensible area within a combinational circuit. In deep sub-micron CMOS devices, the most sensible areas are depletion regions at transistor drains [21]. The particle strike produces several hole-electron pairs that start to drift under the effect of the electric field. As a result, the injected charge tends to change the state of the struck node with a short voltage pulse. As the depletion region is reformed, the charge-drift process decays, and the expected voltage level at the struck node is restored.

In deep sub-micron circuits the capacitance associated to circuit nodes is very small, therefore non-negligible disturbances can be originated even by small amounts of deposited charge, i.e., when energized particles strike the circuit. Considering a typical deposited charge of 3 pC and a node capacitance of 4 pF, we have that the largest possible voltage disturbance is 0.75 Volt [21]. In old 5 Volt CMOS technologies, the magnitude of the voltage swing associated to a SET is about 15% of the normal voltage swing of the node and thus its impact is quite limited, in terms of both duration and magnitude. Conversely, if the technology is scaled to a 3.3 Volt one, the disturbance becomes 22% of a normal swing and thus the transistor that must restore the correct value of the struck node will employ more time to suppress the charge-drift process. Given the considered figures of deposited charge and node capacitance, SET effects on a 1.8 Volt technology will be certainly critical [21]. In very deep sub-micron technologies SET effect may become a critical issue since the duration of the SET-induced voltage pulse may become comparable to the gate propagation delay and thus the voltage pulse may spread throughout the circuit, possibly reaching its outputs. Two consequences may be produced:

- The affected outputs control the clock or the asynchronous reset/preset signals of a number of flip-flops. As a result, the SET is immediately latched by the affected memory elements that change their state.
- The affected outputs are sampled by memory elements thus provoking effects similar to those of SEUs. As described in section 4.7, the latching of a SET depends by several factors.

As measurements reported in [21] show, SET can be conveniently modeled at the gate level as erroneous transitions (either from 0 to 1 or from 1 to 0) on the output of combinational gates. An example of SET is depicted in Fig. *1-15*.

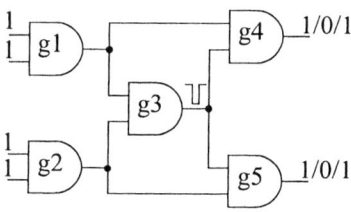

Figure 1-15. An example of Single Event Transient

The circuit primary inputs are set to 1, thus the expected output value is 1 on both g4 and g5 outputs. When g3 is struck by a particle with sufficient energy, its output switches to 0 for a period of time long enough for the spurious transition to propagate through the outputs gates. As a result, we observe a transition on both g4 and g5, whose outputs are set to 0. As soon as the SET effects disappear, the outputs switch back to the expected value.

4.7   Propagating and latching of SETs

Following the terminology introduced in [22], the occurrence, propagation and latch of a SET on a node n to a latch l depends on three factors:

- $R_{SEE}(n)$, which is the probability that a single event having enough energy to produce a SET (which is compatible with the circuit voltage swing) affects the node n. This depends on the device characteristics of the gate driving node n, the amount of capacitance at node n, as well as the sensitive area of node n.
- $P_{sensitized}(n,l)$, which is the probability that at least one path in the circuit from node n to latch l is sensitized, i.e., the SET is free to propagate from the struck node to the latch without being blocked. Whether or not node n is sensitized to latch l depends on the input pattern being applied. Thus, the probability that node n is sensitized to latch l depends on the probability of each input pattern being applied to the circuit while it is operating.
- $P_{latched}(n,l)$, which is the probability that the SET on n is lacthed by l. In order to be captured in latch l, the SET must arrive at the latch during the latching-window in time. The probability of the pulse being present during the latching-window depends on the width of the SET relative to the clock period, and therefore on the amount of the particle's energy.

The sensitivity of a node n with respect to latch l can be expressed as the product of the above terms:

$$Sensitivity(n,l) = R_{SEE}(n) \cdot P_{sensitized}(n,l) \cdot P_{latched}(n,l) \qquad (1)$$

5.   REDUNDANCY TECHNIQUES

All the available techniques to cope with the detection and possibly correction of errors are based on adding to the system some functionalities that are not strictly needed for satisfying the user whishes, i.e., the added functionalities are not involved in carrying out the duties the user demands to the system. The added functionalities' only purpose is to guarantee that any error affecting the system will not harm the system's user, and they will take care of guaranteeing that the system continues to work at least safely if not correctly.

The term that is used to identify the functionalities added to these purposes is *redundancy*, and when used it usually implies the addition of information, resources or time to the system beyond that is needed for normal system operations [3].

Before proceeding with a discussion of the different types of redundancy one can adopt while providing to a system the capabilities for detecting, and correcting possible errors, it is important to remark that redundancy always implies additional costs. Redundancy is not used to implement the operations the system is supposed to perform; conversely, redundancy is used to guarantee that the intended system's functions are performed safely, or correctly even in the presence of errors that may, or may not happen. This implies that, when the occurrence of errors has to be taken into account, the system's user has to pay some extra costs.

In case hardware or information redundancy is used, the user has to pay an extra cost consisting in additional hardware resources that are needed to implement the system.

In case time redundancy is used, the extra cost consists in additional time needed for carrying out the operations the system performs.

5.1   Hardware redundancy

Hardware redundancy consists in the physical replication of the hardware components of a system. Three approaches have been proposed to implement hardware redundancy:

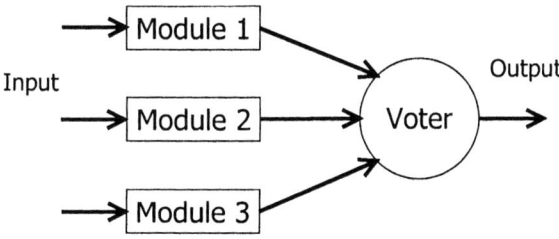

Figure 1-16. The concept of Triple Modular Redundancy

- *Passive redundancy*, which relies upon a voting mechanism to mask the occurrence of errors in a system. A conceptual representation of passive redundancy is presented in Fig. *1-16*, where three identical versions of the system that needs to be protected against errors are connected to a majority voter. This basic concept, known as *Triple Modular Redundancy* (TMR), exploit a majority voter to decide the system's output on the basis of the outputs produced by three identical modules. It one of the module is faulty, the majority voter is still able to decide the system's output by relying upon the two fault-free modules. Passive redundancy is normally used to provide tolerance to errors: since the voter masks errors, they never reach the system's output, which is always correct.

- *Active redundancy* splits the problem of tolerating errors in three phases: *error detection*, *error location*, and *error recovery*. The major difference with respect to passive redundancy is that active redundancy does not try to mask errors. This implies that the output of the system may be erroneous while the system is trying to detect, locate and correct the error. An example of active redundancy is the approach known as *standby sparing*: the system is composed of one operating module, and one or more spare modules. As soon as an error has been detected and localized in the operating module (no matter which fault detection, and location approach is used) the operating module is replaced with one of the spares. The switching between the faulty operating module and one of the fault-free spares implements the recovery phase needed to restore the correct operations. Sparing can be either cold or hot. In *cold standby sparing* spares are idle, and the selected spare is powered up only when it is needed to replace the faulty operational module. During the switching, the service delivered by the system is momentarily disrupted. In case the recovery time needs to be minimized the *hot standby sparing* can be exploited. According to this approach, the spares are powered up and work in parallel to the operating module. As soon as the operating module produces an error, one of the spares can immediately replace it.
- *Hybrid redundancy* combines passive and active redundancy. Error masking is used to inhibit the system to produce erroneous output, while error detection, location and recovery are used to restore the faulty module to a fault-free state.

5.2   Information redundancy

Information redundancy consists in adding redundant information to a data to allow error detection, masking and possibly tolerance [3]. Information redundancy is based on the concept of *code*, which is a mean to represent data using a self-defined set of rules. A piece of data represented according to the rules of a code is known as *codeword*, which me be *valid* in case it adheres to all the rules the code defines, or *invalid* in case it violates at least one of the code's rules. Given a piece of data, the *encoding operation* translates it in a valid codeword. Conversely, the *decoding operation* translates a codeword in the corresponding piece of data.

By selecting the proper rules, it is possible to define:

- *Error-detecting codes*, which allow detecting the occurrence of errors by forming a codeword in such a way that any error affecting it transforms a valid codeword in an invalid one.
- *Error-correcting codes*, which allow identifying from an invalid codeword the corresponding valid one that was corrupted by an error.

As an example of codes, we can consider the *single-bit parity code*. The code mandates the addition of an extra bit to a binary data in such a way that the resulting codeword has an even number of 1s (even parity) or an odd number of 1s (odd parity). If a codeword with odd parity (in contains an odd number of 1s) is affected by an error changing one of its bits, the parity will become even. As a result, known the type of parity (even or odd), it is possible to perform error detection by simply counting the number of 1s in the codeword.

5.3   Time redundancy

The basic concept of time redundancy consists in performing the same operation two or more times, and to compare the results to detect if an error occurred.

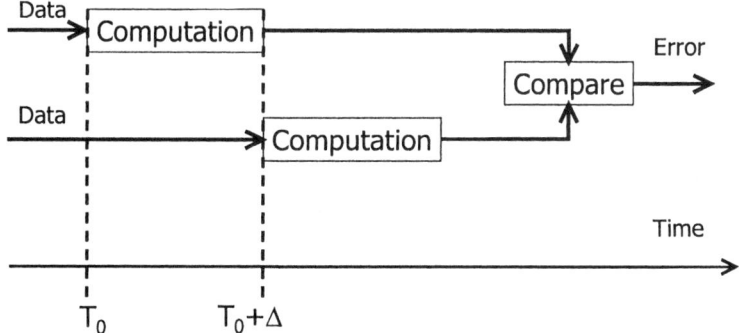

Figure 1-17. Time redundancy for detecting transient errors

In case an error has been detected, the same computation can be repeated again to verify if the error is still present in the system or if it disappeared. Two versions of time redundancy can be envisioned:

- *Time redundancy for transient error detection* is intended for detecting the presence in the system of an error that affected the correct system's operations for a finite period of time. In this case, the scheme of Fig. *1-17* is used, where the same computation is repeated twice, one at time T_0, and one at a later time $T_0+\Delta$. The outcomes of the two computations are compared, and in case a mismatch is found an error is signaled. In order to be effective, the technique relies upon designers to identify a suitable delay Δ between the executions of the two computations in such a way that only one of the two computations is erroneous.

- *Time redundancy for permanent error detection* is an extension of the previous technique, whose aim is to detect permanent errors, i.e., errors that modify the correct system operations for an infinite period of time. The concept at the base of this technique is depicted in Fig *1-18*: the first computation is performed as usual, while before the second computation occurs, the input data are encoded, then elaborated by the computation, and finally the results are decoded and compared with those produced by the first computation. Decode and encode operations are selected in such a way that permanent errors can be detected. Typical operators are complementation and shift [3].

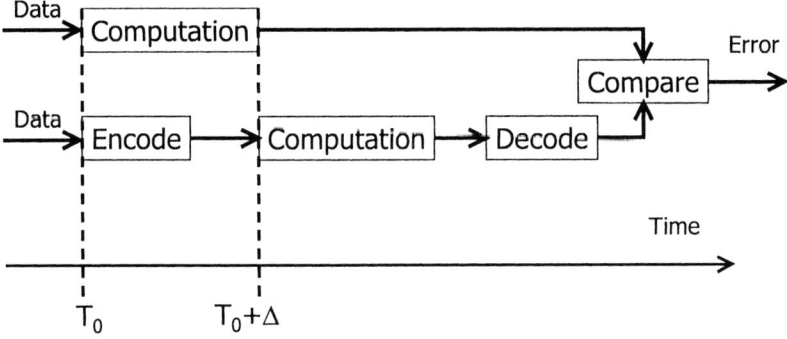

Figure 1-18. Time redundancy for detecting permanent errors

5.4   Software redundancy

Software redundancy is the general term under which the Software-implemented Hardware Fault Tolerance techniques presented in the following chapters falls. Several different approaches have been proposed, which all share the same concepts: additional instructions are added to the original program to implement in software information redundancy and time redundancy.

Since software redundancy is the scope of this book, and it is unfeasible to summarize here all the available techniques, we forward the reader to the following chapters.

6. REFERENCES

1. N. G. Levenson, C. S. Turner, "An investigation of the Therac-25 accidents", IEEE Computer, Vol. 26, No. 7, 1993, pp. 18-41

2. N. G. Leveson, Safeware. System safety and computers. Addison Wesley, ISBN 0-201-11972-2

3. D. K. Pradhan, Fault-tolerant computer system design, Prentice-hall, ISBN 0-13-057887-8

4. J. C. Geffroy, G. Motet, Design of Dependable Computing Systems, Kluwer Academic Publishers, ISBN 1-4020-0437-0

5. A. Avizienis, J.-C. Laprie, B. Randell, C. Lanwehr, "Basic Concepts and Taxonomy of Dependable and Secure Computing", IEEE Transactions on Dependable and Secure Computing, Vol. 1, No. 1, 2004, pp. 11-33

6. N. Storey, Safety-Critical Computer Systems, Pearson/Prentice-Hallp, ISBN 0-201-42787-7

7. M. Abramovici, M. A. Breuer, A. D. Friedman, Digital System Testing and Testable Design, Wiley-IEEE Press, ISBN 0-7803-1062-4

8. E. Dupont, M. Nicolaidis, P. Rohr, "Embedded robustness IPs for transient-error-free ICs", IEEE Design & Test of Computers, Vol. 19, No. 3, May-June 2002, pp. 54-68

9. J. Duraes, H. Madeira, "Emulation of Software Faults by Educated Mutation at Machine-level", IEEE International Symposium on Software Reliability Engineering, 2002, pp. 329-340

10. P. E. Dodd, L. W. Massengill, "Basic Mechanisms and Modeling of Single-Event Upset in Digital Microelectronics", IEEE Transactions on Nuclear Science, Vol. 50, No. 3, June 2004, pp. 583-602

11. J. L. Barth, C. S. Dyer, E. G. Stassinopoulos, "Space, Atmospheric, and Terrestrial Radiation Environments", IEEE Transactions on Nuclear Science, Vol. 50, No. 3, June 2004, pp. 466-482

12. A. H. Taber, E. Normand, "Investigations and characterization of SEU effects and hardening strategies in avionics", Defense Nuclear Agency, Alexandria, VA, DNA-TR-94-123, Feb. 1995

13. A. Holmes-Siedle, L. Adams, Handbook of radiation effects, 2nd edition, Oxford University Press, ISBN 0-19-850733-X

14. J. Barak, J. Levinson, M. Victoria, W. Hajdas, "Direct process in the enrgy deposition of protons in silicon", IEEE Transactions on Nuclear Science, Vol. 43, No. 12, Dec. 1996, pp. 2820-2826

15. S. Duzellier, R. Ecoffet, D. Falguère, T. Nuns, L. Guibert, W. Hajdas, M. C. Calver, "Low energy proton induced SEE in memories", Vol. 44, No. 12, Dec. 1997, pp. 2306-2310

16. T. C. May, M. H. Woods, "Alpha-particle-induced soft errors in dynamic memories", IEEE Transactions on Electronic Devices, Vol. 26, Feb. 1979, pp. 2-9

17. E. Takeda, K. Tacheuhi, D. Hisamoto, T. Toyabe, K .Ohshima, K. Itoh, "A cross section of α-particle-induced soft-error phenomena in VLSIs", IEEE Transactions on Electronic Devices, Vol. 36, Nov. 1989, pp. 2567-2575

18. R. J. McPartland, "Circuit simulations of alpha-particle-induced soft errors in MOS dynamic RAMs", IEEE J. Solid-State Circuits, Vol. 16, Feb. 1981, pp. 31-34

19. T. V. Rajeevakumar, N. Lu, W. Henkels, W. Hwang, R. Franch, "A new failure mode of radiation-induced soft errors in dynamic memories", IEEE Electronic Device Letters, Vol. 9, Dec. 1988, pp. 644-646

20. L. W. Massengill, "Cosmic and terrestrial single-event radiation effects in dynamic random access memories", IEEE Transactions on Nuclear Science, Vol. 43, Apr. 1993, pp. 576-593

21. K. J. Hass, J. W. Gambles, "Single event transients in deep submicron CMOS", IEEE 42nd Midwest Symposium on Circuits and Systems, 1999, pp. 122-125

22. K. Mohanram, N. A. Touba, "Cost-effective approach for reducing soft error failure rate in logic circuits", IEEE International Test Conference, 2003, pp. 893-901

Chapter 2

HARDENING THE DATA

1.   INTRODUCTION

This chapter presents the methods for hardening a system against faults affecting the data it elaborates.

The methods exploit operation and information redundancy and are based on program modifications. The techniques described in the following paragraphs present the following general characteristics (some cases present exceptions emphasized in the specific descriptions):
- The size of the memory area containing the data is at least 2 times the size of the original program.
- The computation time of the resulting program is at least 2 times slower than the original program.
- The programmer has to follow some strict programming rules, concerning the usable data structures and statements.

This means that the adoption of these techniques is rather expensive in terms of memory size, execution slow down, and programming limitations. On the other side, they offer a very good coverage of the addressed faults.

2.   COMPUTATION DUPLICATION

Computations can be duplicated at four levels of granularity: instruction, instructions block, procedure or program.

The smallest granularity is instruction-level, in which an individual instruction is duplicated. For example, the duplicated instruction is executed immediately after the original instruction is executed; the duplicated instruction may perform the same computation carried out by the original instruction, or it can even perform a mutation of the original operation.

The coarsest level of duplication is the program-level, in which the whole program is duplicated: the duplicated program may be executed after the original program completes its execution or it can be executed concurrently.

Whatever the level of granularity is adopted the technique is able to detect faults by executing a check after the duplication is executed. With the instruction-level duplication a check compares the results coming from the original instruction and its duplication; with the procedure-level duplication the results of the duplicated procedures are compared; with the program-level duplication a comparison among the outputs of the programs is executed in order to detect possible faults.

2.1   Methods based on instruction-level duplication

2.1.1   High-level instruction duplication

A simple method to achieve error detection capability is based on introducing data and code redundancy according to a set of transformations to be performed on the high-level source code [23]. The transformed code is able to detect errors affecting both data and code: the goal is achieved by duplicating each variable and adding consistency checks after every read operation. Other transformations focus on errors affecting the code, and correspond from one side to duplicating the code implementing each write operation, and from the other to adding checks for verifying the consistency of the executed operations.

The check operation is executed at every read operation in order to reduce the effect of possible error propagations.

The main advantage of the method lies in the fact that it can be automatically applied to a high-level source code [24], thus freeing the programmer from the burden of guaranteeing its correctness and effectiveness (e.g., by selecting what to duplicate and where to put the checks). The method is completely independent on the underlying hardware, and it possibly complements other already existing error detection mechanisms.

The rules mainly concern the variables defined and used by the program. The method refers to high-level code, only, and does not care whether the variables are stored in the main memory, in a cache, or in a processor register. The proposed rules may complement other Error Detection Mechanisms that can possibly exist in the system (e.g., based on parity bits or on error correction codes stored in memory). It is important to note that the detection capabilities of the rules are significantly high, since they address any error affecting the data, without any limitation on the number of modified bits or on the physical location of the bits themselves.

The basic rules can be formulated as follows:

- Rule #1: every variable x must be duplicated: let x_0 and x_1 be the names of the two copies

- Rule #2: every write operation performed on x must be performed on x_0 and x_1
- Rule #3: after each read operation on x, the two copies x_0 and x_1 must be checked for consistency, and an error detection procedure should be activated if an inconsistency is detected.

The check must be performed immediately after the read operation in order to block the fault effect propagation. Please note that variables should be checked also when they appears in any *expression* used as a condition for branches or loops, thus allowing a detection of errors that corrupt the correct execution flow of the program.

Every fault that occurs in any variable during the program execution can be detected as soon as the variable is the source operand of an instruction, i.e., when the variable is read, thus resulting in minimum error latency, which is approximately equal to the temporal distance between the fault occurrence and the first read operation. Errors affecting variables after their last usage are not detected (but do not provoke any failure, too).

Two simple examples are reported in Fig. *2-1*, which shows the code modification for an *assignment* operation and for a *sum* operation involving three variables *a*, *b* and *c*.

Original code	Modified Code

`a = b;`	`a0 = b0;`
	`a1 = b1;`
	`if (b0 != b1)`
	`    error ();`

`a = b + c;`	`a0 = b0 + c0;`
	`a1 = b1 + c1;`
	`if ((b0!=b1) || (c0!=c1))`
	`    error ();`

Figure 2-1. Example of code modification.

The parameters passed to a procedure, as well as the returned values, should be considered as variables. Therefore, the rules defined above can be extended as follows:

- every procedure parameter is duplicated
- each time the procedure reads a parameter, it checks the two copies for consistency
- the return value is also duplicated (in C, this means that the addresses of the two copies are passed as parameters to the called procedure).

Fig. *2-2* reports an example of application of Rules #1 to #3 to the parameters of a procedure.

Original code	Modified code

`res = search (a);`
 	
...
 	
`int search (int p)`
`{  int q;`
 	
...
 	
`    q = p + 1;`
...
`    return(1);`
`}`
	`search(a_0, a_1, &res_0, &res_1);`
	
	...
	
	`void search (int p_0, int p_1, int *r_0, int *r_1)`
	`{ int q_0, q_1;`
	
	...
	
	`    q_0 = p_0 + 1;`
	`    q_1 = p_1 + 1;`
	`    if (p_0 != p_1)`
	`          error ();`
	
	...
	
	`    *r_0 = 1;`
	`    *r_1 = 1;`
	`    return;`
	`}`

Figure 2-2. Example of code transformation for errors affecting procedure parameters.

In order to assess the effectiveness of the proposed transformation rules, a set of fault injection campaigns has been reported in [25]. They have been performed on a prototypical board (called *Transputer board*) which has been originally designed for carrying out the injection of transient faults.

The Transputer board mainly includes:

- a *T225 Transputer* (a reduced instruction set microprocessor with parallel capabilities). The T225 is the main core of the board, being in charge of all the operations related with data transfer to/from the user and the implementation of test programs;
- a 4 Kbyte *PROM*, containing the executable code of the programs related with the operation of the board (boot, result transfer, program loading)
- a 32 Kbyte *SRAM*, used for the storage of T225 program workspaces, programs and data. The last 2 Kbytes are reserved to data transfer to/from the user;
- an *anti-latchup circuit*, for the detection of abnormal power consumption situations and the activation of the corresponding recovering mechanisms;
- a *watch-dog system*, refreshed every 1.5 seconds by the T225, which has been included in order to avoid system crashes due to events arising on critical targets such as the T225 internal memory cells (registers or flip-flops) or the external SRAM memory areas associated to the program modules (process workspaces).

The board can easily support fault injection experiments. Faults are randomly injected in the proper locations during the program execution. To be consistent with the characteristics of transient errors, the injection of single faults has been performed on randomly selected bits belonging to the code and data area. The injection mechanism is implemented by a dedicated process, which runs in parallel with the tested program. The two programs (the injection program and the program under test) are loaded in the prototype board memory and launched simultaneously. The injection program waits for a random duration, then chooses a random address and a random bit in the memory area used by the program under test and inverts its value. After each injection, the behavior of the program is monitored, the fault is classified, and the results are sent to the PC acting as a host system.

The performed experiments are based on carrying out extensive fault injection sessions on three benchmark programs:

- *Matrix*: multiplication of two 10x10 matrices composed of integer values
- *BubbleSort*: an implementation of the bubble sort algorithm, run on a vector of 10 integer elements
- *QuickSort*: a recursive implementation of the quick sort algorithm, run on a vector of 10 integer elements.

For each benchmark two fault injection sessions have been executed: one on the original version of the program, the other on the modified one. Faults are injected in the memory area containing the program data. The number of faults injected in each session is 1,000 for the original and the modified versions of the program.

Faults were classified according to the categories already presented in Chapter 1.

Obviously, the goal of any fault detection mechanism is to minimize the number of faults belonging to the last category.

Table 2-1. Results of Injecting Faults in the Data Area

	Version	Effect-Less	Software detected	Failure
Matrix	Original	199	0	801
	Modified	188	812	0
Bubble Sort	Original	235	0	765
	Modified	259	741	0
Quick Sort	Original	240	0	760
	Modified	236	764	0

Table *2-1* reports the results of fault injection experiments performed on the memory area containing the data.

Note that for the original program an average percentage of 77% of faults injected in data areas led to wrong program results; on the other hand, considering the modified program, an almost equivalent average percentage of 77% of faults are detected by the software detection mechanism and there are no faults injected in the program data that provoke failures.

Experimental results reporting average area and performance overheads for the above mentioned programs are given in [26] and are shown in Table *2-2*.

Table 2-2. Area and performance overheads with duplication and check hardening approach

Code Segment Size increase	Data Segment Size increase	Executable Code size increase	Performance Slow-down
3.64	2.0	3.4	2.92

2.1.2   Selective instruction duplication

The previous approach presents high levels of fault coverage at a cost of high memory and performance overhead. A selection of the duplicated variables and instructions can be defined in order to tune the trade-off between the level of dependability improvement and the performance degradation due to the code modification.

Reliable Code Compiler (RECCO) [27] supports the designer in identifying both the most critical portions of the code and its most critical variables, suggesting the best modifications towards a safer code. RECCO operates through the following three phases:

- *Code Reliability Analysis*: For each variable a *reliability-weight* is computed, which takes into account the variable *lifetime* and its *functional dependencies* with other variables.

The *life period* of a variable is defined as the period starting from a write operation and ending with the last read operation on the same data preceding the next write operation or the end of the program execution. Fig. *2-3* reports a graphical representation of the life period where a_1, a_2, ..., a_n corresponds to the time instants when a given variable is accessed and w represents a write operation and r a read operation.

The *lifetime* is defined as the sum of all the variable life periods. Data stored in variables with higher lifetime have higher probability of being corrupted, since they are stored in memory for a longer period of time. RECCO performs a static analysis of the code and evaluates the life period parameter as the number of lines of code between the write and the read operation.

A variable v is *descendent* of a given variable w if it is written with the result of an expression which includes w. Variables with a lot of descendent represent a potential criticality for the system: faulty data stored in them are propagated to a large set of other variables. RECCO computes the list of descendents for each variable, analyzing the whole program and building the correspondent *Variable Dependencies Graph* (VDG). VDG is a direct graph, in which nodes represent variables and direct edges represent variable dependencies, as shown in an example in *Fig. 2-4*.

The reliability weight is computed assigning to each variable a linear function of the two parameters (lifetime and functional dependencies). RECCO sorts all the variables according to their reliability weights.

- *Code Re-ordering Phase*: RECCO modifies the original code and generate a more reliable one, functionally equivalent to the original one, but improved in terms of dependability characteristics. The adopted approach consists in performing local optimization aiming at reducing the reliability weight of the variables identified during the Code Reliability Analysis. RECCO applies the code re-ordering technique on portions of code named domains. No read/write dependencies exists among operations belonging to the same domain, i.e., inside a domain no operation reads/writes a variable that is written/read by another operation in the same domain. Therefore, within a domain all the operations can be freely re-ordered without affecting the global program behavior. Inside a given domain, each operation is labeled with a reliability weight, which is a function of the reliability weights of the involved variables. The operations are sorted for decreasing reliability weights and then rescheduled inside the domain itself, in order to minimize the whole reliability weight.

- *Variable Duplication Phase*: RECCO introduces ad-hoc modifications through the variable duplication phase, consisting in coupling some of the variables with shadow variables. The original and the shadow variable behave in the same way, storing the same type of data and being updated, with the same values, at the same time. Periodically monitoring the consistency between the two copies of the variables, it is possible to detect the occurrence of faults in one of the two replicas of the data. Variables coupled with a shadow variable are therefore reliable variables.

RECCO allows the user to trade-off between code reliability level and performance degradation, appropriately setting the reliability requirements: e.g., the user specifies the percentage of variables to be duplicated, and RECCO selects, among all the variables, the ones that are more critical for the application safety.

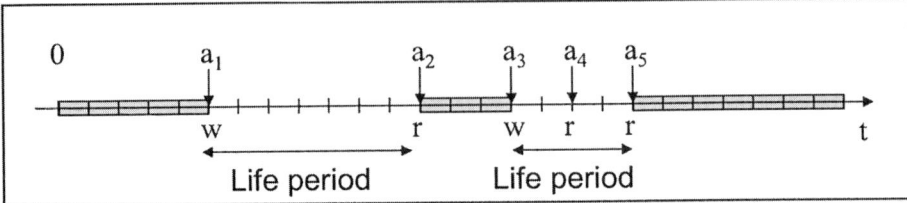

Figure 2-3. Variable's lifetime definition.

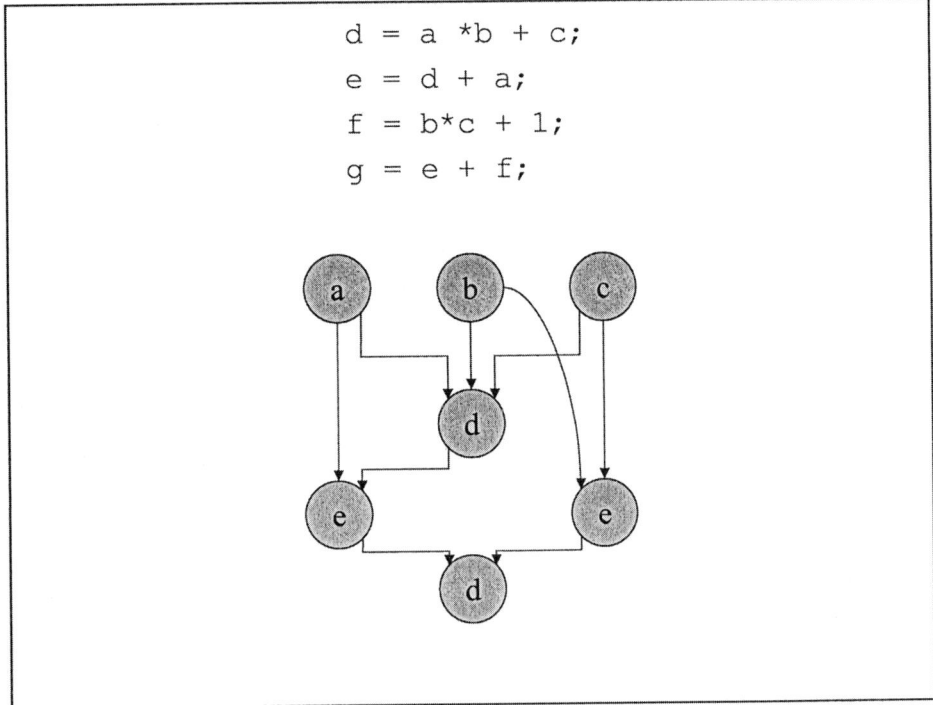

```
d = a *b + c;
e = d + a;
f = b*c + 1;
g = e + f;
```

Figure 2-4. Variable Dependencies Graph.

Experimental results gathered through fault injection experiments on a set of benchmark programs demonstrate that duplicating 30% of the variables, the failures are reduced by 68% with respect to the original code; duplicating the 70% of the variables allows to reach a reduction of failures of 70%. Performance degradation and memory overhead depends strictly on the percentage of variables duplicated: performance slow down by 6% is observed with 30% of variables duplicated; while 18% of memory overhead is needed with 30% of variables duplicated.

2.1.3 Assembly-Level Instruction Duplication

Trends in processor architecture have shown an increasing use of *Instruction-Level Parallelism* (ILP) to improve performance. In addition to pipelining individual instructions, it has become very attractive to fetch multiple instructions at the same time, and execute them in parallel to use functional units whenever possible. This form of ILP is called super-scalar execution. It provides a way to exploit available hardware resources in the system. When superscalar processors are used, it is possible to exploit ILP for error detection.

The basic idea presented by the EDDI (*Error-Detection by Duplicated Instructions*) technique [28] is to duplicate the original instructions in the original assembly source code using duplicated registers and variables, too, according to the following basic rules:

- A *master instruction* (MI) is the original instruction in the source code.
- A *shadow instruction* (SI) is the duplicated instruction added to the source code.
- General purpose registers and memory are partitioned into two groups for MI and SI instructions.
- The registers and memory for MI instructions should always have the same values as the corresponding registers and memory for SI instructions. If there has been a mismatch between a pair of registers for MI and SI, an error can be detected by comparing the values stored into the two registers. A *compare instruction* (CI) compares the values of the two registers, and invokes an error handler if they do not match.

A simple example of source code containing just one MI instruction is the following:

```
        ADD R3, R1, R2              ; R3 <- R1 + R2
```
The corresponding SI and CI instructions can be the following:
```
        ADD R23, R21, R22          ; SI
        BNE R3, R23, gotoError     ; CI
```

Let registers R1, R2, R3 be the master registers, and R21, R22 and R23 the shadow registers that contain the same value as R1, R2, R3, respectively. The CI instruction is executed comparing the values stored in the registers containing the result of the sum (R3 and R23), and if a mismatch is found the control is transferred to an error handler (labeled `gotoError`).

The description of the method requires the following preliminary definitions. A *store* instruction is an instruction that stores the value of a variable in memory. According to this definition, a *Storeless Basic Block*

(SBB) is a sequence of instructions in which there is no store instruction except for the last one, which can be a store, or a branch instruction. An example of SBB is shown in Fig. *2-5*.

```
ADD   R1,  R2,  R3
SUB   R4,  R1,  R2
AND   R5,  R1,  R2
MUL   R6,  R4,  R5
ST    R6
```

Figure 2-5. Example of SBB.

Within a SBB, the SI instructions are scheduled to maximize resource use by attempting to use idle resources, which are not used by MI instructions. A detailed description of the scheduling algorithm is not under the scope of this book, and is presented in details in [28].

If the last instruction of an SBB is a store instruction, then a CI instruction is placed before the store instruction to compare the master and shadow values that are going to be stored in memory.

The EDDI method has been experimentally evaluated on a SGI Octane, that uses the 4-way super-scalar R10000 MIPS processor. Eight benchmark programs were used: FFT, matrix multiplication, Fibonacci, Hanoi, compress, shuffle, Quick sort and Insert sort. The method has been evaluated through a fault injection approach that forces 1 bit-flip in the code segment of the machine code. The location of the bit-flip is determined randomly for each iteration.

On average, in the original programs, 20% of the injected faults produced incorrect outputs and were not detected. On the other hand, only 1,5% of the injected faults in a program hardened with EDDI produce incorrect outputs and were not detected.

Because extra instructions are added to the original assembly code, the program with EDDI suffers from an increase in code size and loss of performance. The execution-time overhead depends on the parallelism available, too, and varies from 13% to 105%. The size overhead strictly depends on the program type, too, and varies from 44% to 113%. The authors made a comparison with [23] and showed that EDDI presents a better error detection capability, thanks to its assembly-level application. EDDI presents a finer grain error detection capability and lower latency. Consequently, it has higher chance of detecting faults that might cause cases of undetected errors that can propagate or get masked.

Detection capability can be obtained exploiting idle cycles of a fine-grain parallel architecture composed of multiple pipelined functional units, where each functional unit is capable of accepting an instruction in every clock cycle. This approach is called *instruction re-execution* [29] and addresses the performance degradation caused by time redundancy. With instruction re-execution, the program is not explicitly duplicated. Rather, when an instruction reaches the execution stage of the processor pipeline, two copies of the instruction are formed and issued to the execution units. Since instructions are duplicated within the processor itself, the processor has flexible control over the scheduling of redundant computations. Dynamic scheduling logic combined with a highly parallel execution core allows the processor to exploit idle execution cycles and execution units to perform the redundant computations. This is possible because there are not always enough independent operations in the program to fully utilize the parallel resources.

A further strategy exploiting parallelism are Very Long Instruction Word (VLIW) processors, that are becoming popular for their ability to process more than one operation per clock cycle. The intrinsic redundancy of the data path units in VLIW processor architectures provides the resources for executing the detection capability concurrently with respect to the *nominal* program (i.e., the original unhardened program).

The insertion of redundant operations for fault detection directly in the source code is not a viable approach since optimization policies tend to detect such redundancy and collapse the original and added operations into a single one, this making the modification useless. Furthermore, by acting on the compiled code, specific optimizations can be performed to minimize code growth and performance degradation.

The approach proposed by Bolchini [30] proposes a second flow of operations created and executed concurrently. This approach considers faults in the register files, thus covering faults in the processor data-path.

Duplication and comparison is the adopted redundancy scheme for achieving the desired hardware fault detection properties. Each operation concerning data path functional units is executed twice and compared in order to detect possible mismatches. The method does not provide the straightforward execution of the same operation on two different functional units, but create a similar operation on a copy of the data stored in the local memory unit. Once data are loaded from main memory, a copy is made and a parallel flow of operations is carried out on the copy, concurrently with respect to original values. A comparison of the produced values is then performed via additional software operations.

The method is based on the implementation of three key elements:
• The compiled application source code (*nominal*).

- A parallel flow of computation on a copy of the nominal values (*checking*).
- Additional operations for comparing corresponding nominal and checking results (*checker*).

The proposed approach that maintains the original hardware architecture consists in compiling the application source code on a reference architecture with one half of the hardware resources and VLIW width. This solution is transparent to both user and system. The checking and checker operations are introduced to fill the unused VLIW word and use the remaining resources. The checking code performs the same operation as the nominal one on a different subset of the register files and on different functional units. The parallel checking code is generated according to the kind of operation.

This approach provides an initial scheduling of the application code on an architecture that has one half of the actual hardware resources, but the experimental results showed the performance degradation ranges from 2% to 25% with respect to the nominal application code. The limited impact can be related to the low average number of operations per clock cycle, which leaves several empty space for duplicated operations.

Full duplication may cause an unacceptable overhead in terms of performance and energy consumption. This is particularly true for large segments of embedded markets where performance and power will continue to be as important as dependability. The approach proposed in [31] presents a technique that fills empty execution slots with duplicate instructions under a performance bound. The compiler determines the instruction schedule by balancing the permissible performance degradation with the required degree of duplication. The objective is to maximize the number of duplicated instructions with a fixed performance overhead. The algorithm considers for each instruction i its *duplication range* that is the range of cycles within which its duplication can be scheduled. This range is determined by the instructions that i depends on as well as the instructions that overwrite the register read by i. The duplicated instruction cannot be scheduled before the source operands for the instruction are read. The algorithm considers each instruction in turn, identifies its duplication range, and creates a duplicate for it if the duplication does not exceed the schedule length by a fixed limit. The experimental results reported figure out that the full duplication incurs an average increase of 42% in the original schedule length, while the method is able to duplicate more than 40% of the instructions without an increase in the original schedule cycles. The percentage of duplicated instructions increases as the performance bound is relaxed. As a consequence, this approach allows the designer to conduct tradeoff analyses between performance and dependability.

2.2   Procedure-level duplication

2.2.1   Selective Procedure Call

The *Selective Procedure Call Duplication* (SPCD) [32] technique is based on the duplication of the procedure execution. The major goals of this approach are the improvement of the system reliability by detecting transient errors in hardware, taking into account the reduction of the energy consumption and of the overhead.

Some industrial experimental results show that significant energy is consumed in clock circuitry and in caches. Therefore, reducing the number of clock cycles and cache access as well as memory access is important to reduce energy dissipation in the system.

SPCD minimizes energy dissipation by reducing the number of clock cycles, cache accesses, and memory accesses by selectively duplicating procedure calls instead of duplicating every instruction. The number of additional clock cycles is reduced because the number of comparisons is reduced by checking the computation results after the original and duplicated procedure execution, instead of checking the results immediately after executing every duplicated instruction. The code size is reduced because some of the procedures are not duplicated. If the code size is reduced the probability of an instruction cache miss can be lowered and energy consumption can be reduced for fetching instructions from the cache to the processor, or moving instructions from the memory to the cache. Also, reducing the number of comparisons decreases the number of data accesses to the data cache and the memory, resulting in reduced energy consumption.

However, there is a trade-off between energy saving and error detection latency: longer error detection latency reduces the number of comparisons inside the procedure and, therefore, saves energy. The shortest error detection latency can be achieved by instruction-level duplication. In procedure-level duplication, comparison of the results is postponed until after executing the called procedure twice; then, the worst case error detection latency corresponds to the execution time of the original and duplicated procedure and the comparison time.

A procedure is a sequence of statements, with an identifying name, executed as a unit through its call in any part of the program. Fig. *2-6* shows the original sample source code where procedure A calls procedure B.

```
int a,c;
void A ()
{
        a = B(b);
        c = c + a;
}
int B (int b)
{
int d;
        d = 2 * b;
        return(d);
}
```

Figure 2-6. Sample source code.

```
int a, a1, c, c1;
void A2 ()
{
        a = B2(b, b1);
        a1 = a;
        c = c + a;
        c1 = c1 + a1;
        if (c <> c1) errorHandler();
}
int B2(int b, b1)
{
int d, d1;
        d = 2 * b;
        d1 = 2 * b1;
        if (d <> d1 ) errorHandler();
        return(d);
}
```

Figure 2-7. Instruction-level duplication.

With instruction-level duplication, all the instructions in the procedures A and B are duplicated as reported in Fig. *2-7*. The code size of the procedures A2 and B2, including comparison statements, is more than twice the original code size of A and B.

A procedure-level duplication is obtained calling twice the procedure; the procedure is called with the original parameter first and then with the duplicated variable as a parameter. Fig. *2-8* shows the resulting source code: the code size of procedure A2 (containing the duplication of the called procedure B) is more than twice the original code size of A, but the size of procedure B is the same in the original and in the modified programs. As a consequence, in a procedure-level duplication program the resulting code size is lower than in an instruction-level duplication one.

```
int a, a1, c, c1;
void A2 ()
{
      a = B(b);
      a1 = B(b1);
      if (a <> a1) errorHandler();
      c = c + a;
      c1 = c1 + a1;
      if (c <> c1) errorHandler();
}
int B(int b)
{
int d;
      d = 2 * b;
      return(d);
}
```

Figure 2-8. Procedure-level duplication

If the called procedure modifies a global variable the duplicated execution of the procedure can introduce an incorrect behavior. Let consider the example shown in Fig. *2-9*, where the procedure B updates the values stored in the global variable *g*.

```
int a,c;
int g;
void A ()
{
     a = B(b);
     c = c + a;
}
int B (int b)
{
int d;
     d = 2 * b;
     g = g + 1;
     return(d);
}
```

Figure 2-9. Sample source code with a global variable modified by the called procedure.

If the procedure B is executed twice, the global variable called *g* is increased twice instead of once. In this case, as shown in Fig. *2-10*, one needs to duplicate the global variable *g1* and the duplicate procedure B1 that modifies this *g1*. The procedures B and B1 are functionally identical, except that B modifies *g* and B1 modifies *g1*.

The basic rules to be considered in a procedure-level duplication approach are the following:

- Every procedure should either be repeated twice or contain duplicated instructions. A procedure that has duplicated instructions can detect an error. A procedure that does not have duplicated instructions should be executed twice, so that an error can be detected.
- If a procedure has no duplicated instructions, all the procedures called by it should have no duplicated statements.

SPCD presents an heuristic algorithm developed to satisfy the previous rules and involving 2 objectives: reducing error detection latency and minimizing energy consumption. In particular, the algorithm presented in [32] minimizes energy consumption under a given error detection latency constraint.

SPCD was simulated with some benchmark programs.

Fault injection experiments were executed injecting single-bit flip faults in the adder unit. Experimental results show that:

- As the error detection latency increases, the energy consumption is reduced

- The data integrity (i.e., the correctness of the outputs) reported is always 100%
- The number of detected faults decreases as the error detection latency increases, but the undetected faults don't cause any failure because they don't affect the final results.

In order to evaluate the feasibility of the approach in terms of energy consumption saving, SPCD is compared with the hardened program obtained applying an instruction-level duplication approach [28]. The obtained results show that SPCD allows an energy saving of 25% with respect than the energy consumption required by an instruction-level duplication approach.

```
int a, a1, c, c1;
int g, g1;
void A2 ()
{
        a = B (b);
        a1 = B1 (b1)
        if (a <> a1) errorHandler ();
        c = c + a;
        c1 = c1 + a;
        if (c <> c1) errorHandler ();
}
int B (int b)
{
int d;
        d = 2 * b;
        g = g + 1;
        return (d);
}
int B1 (int b)
{
int d;
        d = 2 * b;
        g1 = g1 + 1;
        return (d);
}
```

Figure 2-10. Sample source code with a duplicated global variable modified in the called procedure.

2.3   Program-level duplication

2.3.1   Time redundancy

Time redundancy is a technique in which a computation is performed multiple times on the same hardware. A particular application of time redundancy is the *duplication* of the processing activity as a proper technique to detect faults of the underlying hardware. A particular form of such duplication is a virtual duplex system (VDS), where the duplicity is achieved by temporal redundancy, obtained by executing two programs performing the same task with the same input data twice. Virtual duplex systems provide a cost advantage over duplex systems because of reduced hardware requirements: VDS only needs a single processor, which executes both software variants. Transient hardware errors are covered due to time redundancy, as only a single variant is affected. Permanent hardware errors are covered due to design diversity: the program variants of a VDS are diversified in order to reduce the probability that both variants are affected in the same way.

The disadvantage of time redundancy is the performance degradation caused by repetition of tasks.

There are different kinds of duplication: one option consists in running entire programs twice, whereby another option is to execute the duplicated processes in short rounds and switch between them. The switching introduces extra overhead, but can be used to compare intermediate results more frequently in order to reduce the fault latency.

The structure of a VDS is reported in Fig. *2-11*. Each version of program is called *variant*. A VDS built to calculate a specified function f consists of two diversified program variants P_a and P_b calculating the functions f_a and f_b, respectively. In absence of faults $f = f_a = f_b$ holds. If an existing fault affects only one of the two variants or both of them in different ways, then the fault can be detected comparing the results $f_a(i)$ and $f_b(i)$.

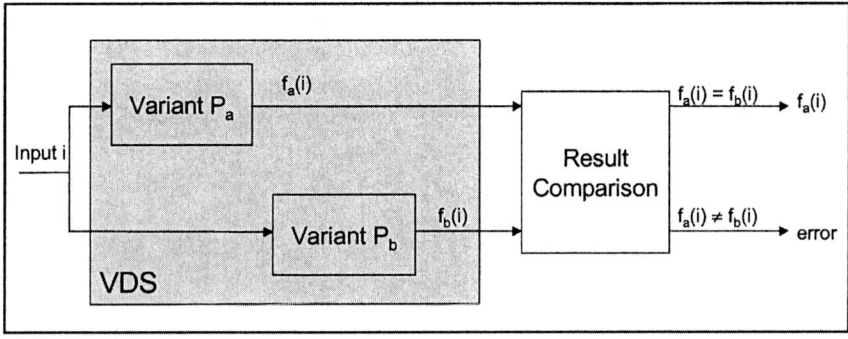

Figure *2-11. Structure of a VDS.*

The kind of faults to be detected by a VDS highly depends on the diversity techniques used to generate the VDS.

As far as VDS is considered, if, for example, two independent teams are developing different variants of a program, then the resulting VDS may have the ability to detect specification or implementation faults. If, as a second example, two different compilers are used to compile the same source code, then the resulting VDS may have the ability to detect faults stemming from compiler faults. The capability of diversified program variants to detect hardware faults has been mentioned and investigated in [33]. The basic idea is that two diversified programs often use different parts of the processor hardware in different ways with different data.

Variants can also be generated by applying manually different diversity techniques. However, some algorithmic approaches have been proposed in order to properly generate effective software variants.

In [38] a systematic method is presented based on the transformation of every instruction of a given program into a modified instruction or sequence of instructions, keeping the algorithm fixed. The transformations are based on a diverse data representation. Since a diverse data representation also requires a modification of instructions that may be executed, new sequences of instruction have to be generated, that calculate the result of the original instruction in the modified representation. The transformations are generated at the assembler and the high-level programming language. Some examples of the modification rules are:

- logical instructions can be modified according to the de Morgan Rules (e.g., a or b = NOT (NOT (A) AND NOT (B)))
- arithmetic instructions can be modified according to the two's complement properties (a+b = - (-a) + (-b))).

In [35] a method for the automated generation of variants is proposed. The tool is able to generate two different but semantically equivalent pieces

of assembler code, exploiting a set of modification rules. Some examples of modification rules are:
- Replacement of jump instructions (e.g., replacement of conditional jump instructions by appropriate combinations of other jump instructions)
- A consistent register permutation
- Substitution of a multiplication statement by a subroutine that performs multiplication in a different way.

2.3.2   Simultaneous multithreading

Simultaneous multithreading (SMT) is a novel technique to improve the performance of a superscalar microprocessor. A SMT machine allows multiple independent threads to execute simultaneously, i.e., in the same cycle, in different functional units. VDS can be effectively exploited on a SMT machine, executing two threads in parallel, shifting time redundancy to spatial redundancy [36]. Because of the improved processor utilization and the absence of a context switch the time execution is reduced with respect to the correspondent duplicated implementation on a conventional processor.

With the *Active-Stream/Redundant-Stream Simultaneous multithreading* (AR-SMT) [37] approach two explicit copies of the program run concurrently on the same processor resources as completely independent programs, each having its own state or program context. The entire pipeline of the processor is conceptually duplicated. As described in Section 2.1.3, in superscalar processors often there are phases of a single program that do not fully utilize the microprocessor architecture, so sharing the processor resources among multiple programs will increase the overall utilization. Improved utilization reduces the total time required to execute all program threads, despite possibly slowing down single thread performance. AR-SMT is based on 2 streams: active stream (A-stream) and redundant instruction stream (R-stream). The active stream corresponds to the original program thread and as instructions from the A-stream are fetched and executed, and their results committed to the program's state, the results of each instruction are also pushed on a FIFO queue called *Delay Buffer*. Results include modifications to the Program Counter by branches and any modifications to both registers and memory. The second stream (R-stream) is executed simultaneously with the A-stream. As the R-stream is fetched and executed, its committed results are compared to those stored in the Delay Buffer. A fault is detected if the comparison fails, and the committed state of the R-stream can be used as a checkpoint for recovery. Simulations made on a processor composed of 8 Processing Elements show that AR-SMT increases execution time by only 10% to 40% over a single thread thanks to the optimized utilization of the highly parallel microprocessor.

2.3.3   Data Diversity

The method exploits data diversity, by executing two different programs with the same functionality, but with different data sets and comparing their outputs. This technique is able to detect both permanent and transient faults.

This approach needs two different programs starting from the original program and transforming it into a new one in which all variables and constants are multiplied by a *diversity factor k*. Depending on the factor k, the original and the transformed programs may use different parts of the underlying hardware and propagate fault effects in different ways. If the two programs produce different outputs due to a fault, the fault can be detected by examining if the results of the transformed program are also k times greater than the results of the original program. The check between the two programs can be executed in two different ways:

1. another concurrent running program compares the results
2. the main program that spawns the original program and the transformed program checks their results after they are completed.

The program transformation changes a program P into a new program P' with diverse data in which all variables and constants are k-multiples of the original values when the program P' is executed. It consists of two transformations:

1. expression transformation
2. branching condition transformation.

The expression transformation changes the expressions in P to new expressions in P' so that the value of every variable or constant in the expression of P' is always the k-multiple of the corresponding value in P. Since the values in P' are different from the original values, when we compare two values in a conditional statement, the inequality relationship may need to be changed if the diversity factor is negative. For example, the conditional statement `if (i<5)` in P needs to be changed to `if (i > - 10)` in P' when $k = -2$.

The branching condition transformation adjusts the inequality relationship in the conditional statement in P' so that the control flows in P and P' are identical.

The sample program in Fig. *2-12* is transformed to a diverse program shown in Fig. *2-13* where $k = -2$.

```
x = 1;
y = 5;
i = 0;
while (i < 5) {
        z = x + i * y;
        i = i + 1;
        }
i = 2 * z;
```

Figure 2-12. Sample program *P*.

```
x = -2;
y = -10;
i = 0;
while (i > -10) {
        z = x + i * y / (-2);
        i = i + (-2);
}
i = (-4) * z / (-2);
```

Figure 2-13. Transformed program *P'*.

The choice of the most suitable value for *k* has to satisfy two goals:
1. to guarantee data integrity, that is, to avoid that two programs produce identical erroneous outputs
2. to maximize the probability that two programs produce different outputs for the same hardware fault in order to achieve error detection.

However, the factor *k* should not cause an overflow in the functional units. The primary cause of the overflow problem in the transformed program is the fact that, after multiplication by *k*, the size of the resulting data may bee too large to fit into the data word size of the processor. For example, consider an integer value of $2^{31} - 1$ in a program (with 32-bit 2's complement integer representation). If the value of *k* is 2, then the resulting integer $(2^{32} - 1)$ cannot be represented using 32-bit 2's complement representation. The overflow problem can be solved by scaling: scaling up to higher precision or scaling down the original data. Scaling up the data to higher precision requires a data type with a larger size. For example, data type such as 16-bit single precision integers can be scaled up to 32-bit double precision integer data type. Scaling up may cause performance overhead because the size of the data is doubled. On the other hand, scaling down the original data (the same effect as dividing the original data by k instead of multiplying by k) will not cause any performance overhead. However, there is a possibility that scaling down data may cause computation inaccuracy during the execution of the program. In this case, when the scaled down values are compared with the original values, only the higher order bits, that are not affected by scaling down, have to be compared.

A first method [38] proposed to consider $k = -1$, i.e., data are complemented.

The method proposed in [39], called ED^{4}I (*Error Detection by Diverse Data and Duplicated Instructions*), demonstrated that, in different functional units, different values of k maximize the fault detection probability and data integrity (for example the bus has the highest fault detection probability when $k = -1$, but the array multiplier has the highest fault detection probability when $k = -4$). Therefore, programs that use a particular functional unit extensively need preferably a certain *diversity factor k*. Considering six benchmark programs (Hanoi, Shuffle, Fibonacci, Lzw compression, Quick sort, Insert sort), the most frequently used functional units are adders and $k = -2$ is the optimum value. On the other hand, the matrix multiplication program extensively uses the multiplier and the optimum value is $k = -4$.

The hardening technique introduces an memory overhead higher than 2 times the memory required for the original program and the performance overhead is higher than 2, too.

ED^{4}I is applicable only to programs containing assignments, arithmetic operations, procedure calls and control flow structures, and cannot applied to statements executing logic operations (e.g., Boolean functions, shift or rotate operations) or exponential or logarithmic functions.

3. EXECUTABLE ASSERTIONS

The method is based on the execution of additional statements that check the validity of the data correspondent to the program variables.

The effectiveness of executable assertions is highly application dependent. In order to develop executable assertions, the developers require extensive knowledge of the system.

Error detection in the form of executable assertions can potentially detect any error in internal data caused by software faults or hardware faults. When input data arrive at a functional block, they are subject to executable assertions determining whether they are acceptable. Output data from computations may also be tested to see if the results seem acceptable.

The approach proposed in [40] describes a rigorous way of classifying the data to be tested. The two main categories in the classification scheme are *continuous* and *discrete* signals. These categories have subcategories that further classify the signal (e.g., the continuous signals can be divided into monotonic and random signals). For every signal class a specific set of constraints is set up, such as boundary values (maximum and minimum values) and rate limitations (minimum and maximum increase or decrease rate), which are then used in the executable assertions. Error detection is performed as a test of the constraints. A violation of a constraint is interpreted as the detection of an error.

Executable Assertion and best effort recovery are proposed in [41], considering a control application. The state variables and outputs are protected by executable assertions to detect errors using the physical constraints of the controlled object. The following erroneous cases can be detected:

- if an incorrect state of the input variable is detected by an executable assertion during one iteration of the control algorithm, a recovery is made by using the state backed-up, during the previous iteration of the computation. This is not a true recovery (as we will see in Chapter 4), since the input variable may differ from the value used in the previous iteration. This may result in the output being slightly different from the fault-free output, thus creating a minor value failure (*best effort recovery*).
- If an incorrect output is detected by an executable assertion, recovery is made by delivering the output produced in the previous iteration. The state variable is also set to the state of the previous iteration that corresponds to the delivered output. This is a best effort recovery, too, since the output could be slightly different from the fault-free value.

Executable assertions with best effort recovery has been experimentally applied on a embedded engine controller [41]. Fault injection experiments executed on the original program showed that 10.7% of the bit-flips injected into data cache and internal register of a CPU caused a failure in the system. Fault injection experiments run on the hardened program modified with the executable assertions with best effort recovery showed that the percentage of failures is decreased to 3.2%, demonstrating that software assertions with best effort recovery can be effective in reducing the number of critical failures for control algorithms.

4. REFERENCES

23. M. Rebaudengo, M. Sonza Reorda, M. Torchiano, M. Violante, "Soft-error Detection through Software Fault-Tolerance techniques", Proceedings of the IEEE International Symposium on Defect and Fault Tolerance in VLSI Systems, 1999, pp. 210-218

24. M. Rebaudengo, M. Sonza Reorda, M. Torchiano, M. Violante, "A source-to-source compiler for generating dependable software", IEEE International Workshop on Source Code Analysis and Manipulation, 2001, pp. 33-42.

25. P. Cheynet, B. Nicolescu, R. Velazco, M. Rebaudengo, M. Sonza Reorda, M. Violante, "Experimentally evaluating an automatic approach for generating safety-critical software with respect to transient errors", IEEE Transactions on Nuclear Science, Vol. 47, No. 6, December 2000, pp. 2231-2236

26. M. Rebaudengo, M. Sonza Reorda, M. Torchiano, M. Violante, "An experimental evaluation of the effectiveness of automatic rule-based transformations for safety-critical applications", IEEE International Symposium on Defect and Fault Tolerance in VLSI Systems, 2000, pp. 257-265

27. A. Benso, S. Chiusano, P. Prinetto, L. Tagliaferri, "A C/C++ source-to-source compiler for dependable applications", IEEE International Conference on Dependable Systems and Networks (DSN), 2000, pp. 71-78.

28. N. Oh, P.P. Shirvani, E.J. McCluskey, "Error Detection by Duplicated Instructions In Super-scalar Processors", IEEE Transactions on Reliability, Vol. 51, No. 1, March 2002, pp. 63-75

29. G. Sohi, M. Franklin, K. Saluja, "A study of time-redundant fault tolerance techniques for high-performance pipelined computers", 19-th International Fault Tolerant Computing Symposium, 1989, pp. 463-443

30. Bolchini, C., "A software methodology for detecting hardware faults in VLIW data paths", IEEE Transactions on Reliability, Vol. 52, No. 4, Dec. 2003, pp. 458-468

31. J.-S. Lu, F. Li, V. Degalahal, M. Kandemir, N. Vijaykrishnan, M.J. Irwin, "Compiler-directed instruction duplication for soft error detection", Proceedings of Design, Automation and Test in Europe, 2005, pp. 1056-1057

32. N. Oh, E. J. McCluskey, "Error Detection by Selective Procedure Call Duplication for Low Energy Consumption", IEEE Transactions on Reliability, Vol. 51, No. 4, December 2002, pp. 392-402

33. K. Echtle, B. Hinz, T. Nikolov, "On Hardware Fault Detection by Diverse Software, Proceedings of the 13-th International Conference on Fault-Tolerant Systems and Diagnostics," 1990, pp. 362-367

34. H. Engel, "Data flow transformations to detect results which are corrupted by hardware faults", Proceedings of IEEE High-Assurance Systems Engineering Workshop, 1996, pp. 279-285

35. M. Jochim, "Detecting processor hardware faults by means of automatically generated virtual duplex systems", Proceedings of the International Conference on Dependable Systems and Networks, 2002, pp. 399 – 408

36. S. K. Reinhardt, S.S. Mukherjee, "Transient Fault Detection via Simultaneous Multithreading," Proceedings of the 27th International Symposium on Computer Architecture, 2000, pp. 25-36

37. E. Rotenberg, "AR-SMT: a microarchitectural approach to fault tolerance in microprocessors", 29-th International Symposium on Fault-Tolerant Computing, 1999, pp. 84-91

38. H. Engel, "Data Flow Transformations to Detect Results which are corrupted by hardware faults", Proceedings of the IEEE High-Assurance System Engineering Workshop, 1997, pp. 279-285

39. N. Oh, S. Mitra, E. J. McCluskey, "ED4I: Error detection by diverse data and duplicated instructions", IEEE Transactions on Computers, Vol. 51, No. 2, February 2002, pp. 180-199

40. M. Hiller, "Executable assertions for detecting data errors in embedded control systems", Proceedings International Conference on Dependable Systems and Networks, 2000, pp. 24-33

41. J. Vinter, J. Aidemark, P. Folkesson, J. Karlsson, "Reducing Critical Failures for Control Algorithms Using Executable Assertions and Best Effort Recovery", Proceedings of the International Conference on Dependable Systems and Networks, 2001, pp. 347-356

Chapter 3

HARDENING THE CONTROL FLOW

1. INTRODUCTION

This chapter presents the main software-implemented techniques for hardening a microprocessor-based system against control flow (CF) errors (CFEs). A CFE is an error that causes a processor to fetch and execute an instruction different than expected.

As experiments demonstrate, a significant percentage of transient faults leads to CFEs: in the experiments performed in [42] in average around 78% of faults affecting a system caused CFEs (of course, this figure depends a lot on the processor architecture and on the applications on which experiments are performed). Most of the CFEs cannot be identified by the mechanisms developed for data errors identification presented in chapter 2. These reasons stimulate the development of special mechanisms for CFEs identification, which are presented in this chapter.

2. BACKGROUND

The program code can be partitioned into basic blocks (BBs). A *BB* (sometimes also named *branch free interval*) of a program is a maximal sequence of consecutive program instructions that, in absence of faults, are always executed altogether from the first one to the last one.

From the definition of a BB it follows that a BB does not contain any instruction that may change the sequential execution, such as jump or call instructions, except for the last one, possibly. Furthermore, no instructions within the BB can be the destination of a branch, jump or call instruction, except for the first one, possibly [42].

A *BB body* is the BB without the last jump instruction. If the last BB instruction is not a jump instruction, then the BB body coincides with the BB. It is possible that a BB body is empty if the BB consists of one jump instruction, only.

0	`i = 0;` `while(i < n) {`
1	`if (a[i] < b[i])`
2	`x[i] = a[i];`
3	`else` `x[i] = b[i];`
4	`i++;` `}`
5	

a)

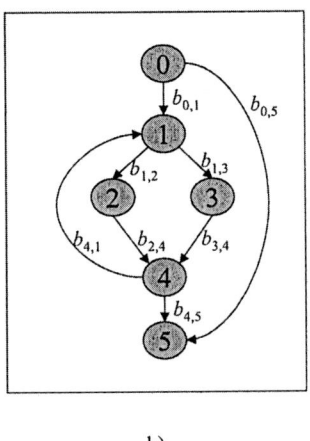

b)

Figure 3-1. Example of program source code and its CFG

A program P can be represented with a CF Graph (CFG) composed of a set of nodes V and a set of edges B, $P = \{V, B\}$, where $V = \{v_1, v_2, ..., v_n\}$ and $B = \{b_{i1,j1}, b_{i2,j2}, ..., b_{im,jm}\}$. The CFG represents the CF of a program. Each node $v_i \in V$ represents a program section, which can be a single instruction or a block of instructions, for example a BB. Each edge $b_{i,j} \in B$

represents the branch from node v_i to node v_j [43]. In the following we will consider CFGs where nodes represent BBs, and use the terms node and BB interchangeably, unless otherwise explicitly stated.

As an example, let us consider the sample program fragment shown in Fig. 3-1 (where the BBs are numbered, and the corresponding program CFG is shown).

Considering the program CFG $P = \{V, B\}$, for each node v_i it is possible to define $suc(v_i)$ as the set of nodes successor of v_i and $pred(v_i)$ as the set of nodes predecessor of v_i [44]. A node v_j belongs to $suc(v_i)$ if and only if $b_{i,j}$ is included in B. Similarly, v_j belongs to $pred(v_i)$ if and only if $b_{j,i}$ is included in B. For example, in the CFG in Fig. 3-1 b) $suc(1) = \{2,3\}$ and $pred(1) = \{0,4\}$.

Let a program be represented by its CFG $P = \{V, B\}$. A branch $b_{i,j}$ is *illegal* for P if $b_{i,j}$ is not included in B [43]. If a program, due to a fault, executes a branch $b_{i,k} \in B$ instead of the correct branch $b_{i,j}$, then the branch $b_{i,k}$ is *wrong*.

Illegal and wrong branches represent CFEs.

In Fig. 3-2 a) and 3-2 b) two examples of CFEs are presented where the error branches are represented with dotted lines. These CFEs can be caused (for example) by faults in the offset operand of the instructions corresponding to branches $b_{2,4}$ and $b_{4,1}$, respectively, which transform them to branches $b_{2,5}$ and $b_{4,5}$, respectively. In Fig. 3-2 a) a branch introduced by a fault is illegal as the set B of the CFG $P = \{V, B\}$ does not contain branch $b_{2,5}$. In Fig. 3-2 b) a branch introduced by a fault is wrong as the set B of the CFG $P = \{V, B\}$ contains this branch.

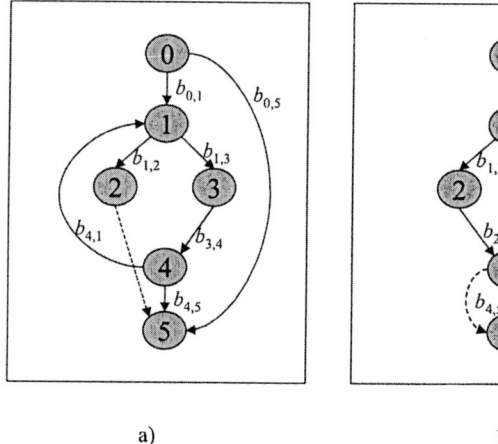

a)   b)

Figure 3-2. Examples of CFEs: a) illegal branch, b) wrong branch

CF checking (CFC) approaches are approaches detecting CFEs. According to the purpose of this book, in this chapter we will refer to purely software CFC approaches, *i.e.*, approaches that do not need any hardware architecture modification for their implementation. Chapter 5 will introduce some methods exploiting some hardware components to achieve the same goal.

CFEs can be divided into *intra-block* errors, which cause erroneous branches having as their source and destination different blocks, and *inter-block* errors, which cause erroneous branches not crossing the blocks boundaries. Correspondingly, intra-block CFC techniques control that instructions inside a block are executed in the correct order, and inter-block CFC techniques detect inter-block CFEs.

Most of the purely software CFC approaches presented in literature are oriented just to inter-block CFC and only a small part of these are oriented to both inter-block and intra-block CFC.

A common approach for the software-implemented detection of CFEs, causing erroneous branches inside program area, is the signature-monitoring technique. In this approach monitoring is performed by regular processor instructions (called monitoring or checking code) embedded into the program under execution. A signature (or identifier) is associated to program structure (it can be a singular instruction, a block of instructions, a path of the program CFG, or other) during compile time or by special program prior to program execution. During program execution a run-time signature is computed. Periodically, checks for consistency between the reference signature and the run-time signature are performed. The mismatch signals the CFE. The run time signature computed during program execution is usually stored in a special area, *e.g.*, a processor register.

The difference among software-implemented CFC approaches mainly consists in the way signatures are computed and checks are performed.

We evaluate CFC approaches presented in this chapter basing on the fault model, which stems from the fault models proposed in literature (*e.g.*, [44], [45], [46]) and includes the following types of CFEs:

Type 1. A fault causing an illegal branch from the end of a BB to the beginning of another BB.

Type 2. A fault causing a legal but wrong branch from the end of a BB to the beginning of another BB.

Type 3. A fault causing a branch from the end of a BB to any point of another BB body.

Type 4. A fault causing a branch from any point of a BB body to any point of different BB body.

Type 5. A fault causing a branch from any point of a BB body to any point in the same BB body.

These types of CFEs are schematically presented in Fig. 3-3. In this figure rectangles denote BB bodies and arrows denote erroneous branches caused by CFEs.

The first 4 types represent inter-block CFEs, while type 5 represents intra-block CFEs. The considered fault model includes only CFEs, which lead to erroneous branches inside the program memory.

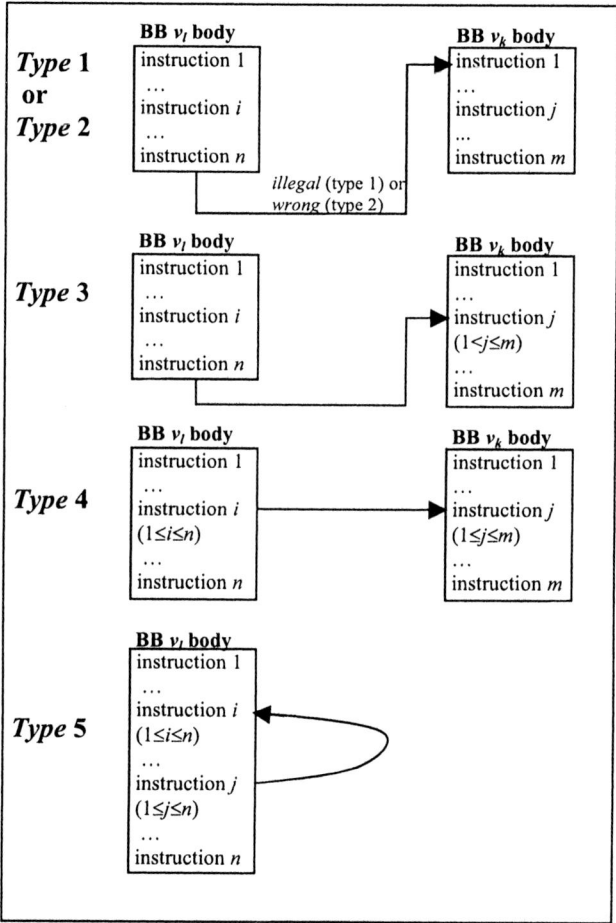

Figure 3-3. Considered types of CFEs

Let now consider the types of CFEs included in the fault model with respect to the system-level errors considered in chapter 1. The presented types of CFEs correspond to single code or data system-level errors, such as:
- errors in the offset of a branch instruction (CFE types 1, 2 and 3),

- errors in a condition upon which a conditional branch is taken (CFE type 2),
- errors changing a nonbranch instruction to a branch one (CFE types 4 and 5),
- errors changing a branch instruction to a nonbranch one (CFE types 1 and 2).

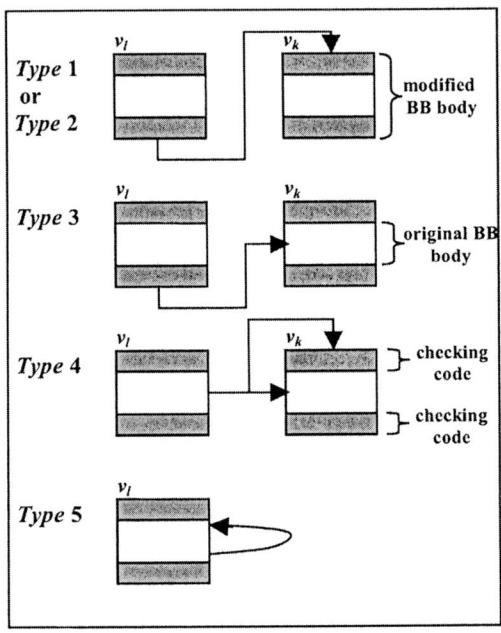

Figure 3-4. Considered types of CFEs for the hardened program

In this chapter we evaluate (not strictly) the capabilities of each CFC method to cover the types of CFEs from introduced fault model; such evaluation is performed in the subsection named "Advantages and limitations". In this evaluation we consider CFEs, which appear in the original code. However, during the program execution the erroneous branches can have as their source and/or destination instructions of the additional checking code. For the purpose of simplicity we do not evaluate the coverage of such CFEs; only for some methods we put remarks on them. However, as these CFEs can lead to erroneous functionality of the program it is important to take them in consideration when the CFC method is developed. In Fig. 3-4 the CFEs types considered during evaluation are graphically presented for modified BBs hardened with checking code. Here it is supposed that the checking code is added in the beginning and/or at the end of each BB, which is the case in most CFC methods.

The probability of erroneous branches having as their source some instructions of the additional checking code increases in case the checking code itself introduces new branches, which can be as well sources of CFEs. That is why it is desirable that the checking code is either branch free or the method is developed taking in consideration these branches.

Among the purely software CFC approaches it is possible to distinguish those which work on assembly-level program code and those which work on high-level program code. The latter ones are more attractive due to their higher portability, since the hardened version of the program is independent from the platform it is intended to be run on. Nevertheless, the high-level approaches have the drawback that the high-level CFG may not correspond exactly to the assembly-level CFG, and this may lead to lower error coverage. For example, in Fig. 3-5a) the high-level C instruction i++ (where i is an integer variable) is presented. If the program containing this instruction is compiled for a processor, which contains an 8-bit ALU, then the increment operation can be performed in two steps (Fig. 3-5 b)): first the lower byte is increased and only if its value becomes zero the higher byte is increased too. So a new jump is introduced by the operation and consequently the high-level program BB containing the instruction i++ is split into several BBs in the assembly-level code, and new branches are introduced in the assembly-level program CFG. Special techniques have been proposed to tackle this problem [47].

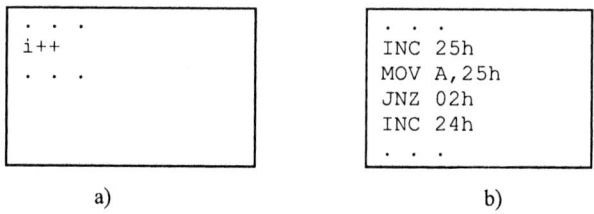

a)   b)

Figure 3-5. A high-level instruction and its assembly-level representation

In the next sections of this chapter the CFC techniques proposed in literature are presented. We put them in chronological order.

For each of the CFC techniques the subsection "The approach" briefly describes the approach presented by its authors in the corresponding papers. The subsection "Experimental results" reports experimental results reported by the authors of the corresponding approach; these results are presented briefly and figures are reported in average (most of the average figures are rounded off); for details the reader should refer to the corresponding papers.

The exception is the experiments performed by the authors of this book and reported in section 13.2, which are presented in more details. The subsection "Advantages and limitations" presents evaluation of the approach performed by the authors of this book.

In sections 4, 5, 6, 7, 11 some assembly-level CFC techniques are presented; sections 3, 8, 9, 10, 12, 13 and 14 describe some high-level CFC techniques.

3.   PATH IDENTIFICATION

3.1   The approach

In this section we summarize the approach presented in [43]. This approach is oriented to the detection of CFEs resulting from both software coding errors and hardware faults.

According to the approach, the program is partitioned into loop-free intervals. A data-base, which contains the paths information for each of the program loop-free intervals, is constructed and the code for the CFEs detection is added to the program. During the program execution for each traversed loop-free interval the traversed path information is recorded and on the next loop-free interval entry it is verified for consistency with the information in the data-base. In the case of discrepancy, a CFE is detected.

The data-base information may be obtained either from the program design or extracted from the code. In the former case the method is also able to detect possible software coding errors.

A *loop-free interval* is defined as a maximal subgraph of the program CFG, which does not contain loops and has a single entry. The partition of the CFG into loop-free intervals has the following properties:
- it is unique for CFG
- it is not complete, as not all branches of the CFG are included into some loop-free interval.

In Fig. 3-6 a) the loop-free intervals of the CFG from Fig. 3-1 b) are presented; here branches $b_{0,1}$, $b_{0,5}$, $b_{4,1}$ and $b_{4,5}$ are not included into any loop-free interval.

A unique prime number called *vertex identifier* is associated to each BB within each loop-free interval; each path in a loop-free interval is represented by a *path identifier*, which is the product of the vertex identifiers of the BBs included in the path. This representation is intended to satisfy the following properties: compactness, uniqueness and unambiguousness. The first two properties are satisfied by the proposed path representation, the third property is satisfied in the case of single CFEs; multiple CFEs in some cases can cause the error compensation or aliasing. A unique identifier (*ID*) number is associated to each loop-free interval of the program.

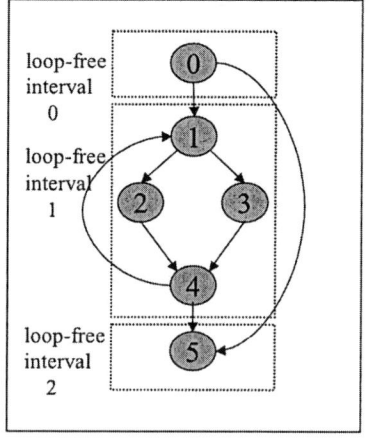
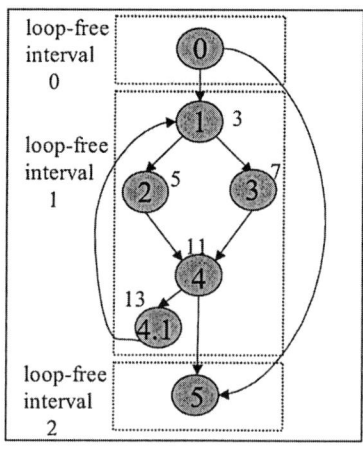

a)   b)

Figure 3-6. Loop-free intervals

For each loop-free interval a path table is constructed, which contains for each path of the loop-free interval a line, containing a current loop-free interval *ID* (*CIID*), a path predicate, a path identifier and the next loop-free interval *ID* (*NIID*).

For the purpose of path table construction each loop-free interval terminal BB with *m* outcoming branches (where *m*>1) *m*-1 dummy BBs are introduced with one outcoming branch. In Fig. 3-6 b) the loop-free interval 1 is prepared to path table construction: a dummy BB 4.1 is introduced and vertex identifiers are assigned to each BB belonging to the loop-free interval 1. The example of the path table corresponding to the loop-free interval 1 for the program from Fig. 3-1 a) is presented in Table 3-1.

Table 3-1. Example of path table

Current Loop-Free Interval CIID	Path Predicate	Path Identifier	NIID
1	a[i]<b[i]; i<n	2145	1
1	a[i] ≥b[i]; i<n	3003	1
1	a[i]<b[i]; i≥n	165	2
1	a[i] ≥b[i]; i≥n	231	2

The program is supplied with the following variables:

- A global variable *NIID*, which contains the *ID* of the next loop-free interval to be traversed. At the program start the *NIID* variable is set to the *ID* of first program loop-free interval.
- a *CIID* variable, which is introduced in each module (*e.g.*, a procedure) of the program and contains the *ID* of the currently traversed loop-free interval.
- a *RPI* variable, which is introduced in each module of the program and contains the identifier of the currently traversed path.
  Some checking code is added to the program to perform run-time CFC.

At the beginning of each loop-free interval the code presented in Fig. 3-7 a) is added. Here, function *TAB* checks for the correctness of the path traversed in the previous loop-free interval. For this purpose it searches the path identifier saved in the *RPI* variable in the previous interval's path table; if it is not found, an error is detected; otherwise, the path predicate is checked using the stored input control variables[1] of a loop-free interval; if it does not correspond to the predicate recorded in the path table, then an error is detected. If the path is correct, TAB function saves the new value for the *NIID* variable taken from the last column of the path table. The TAB function is not added to the first loop-free interval of the program or of some module.

```
TAB(CIID,RPI,NIID);
if(NIID!=ID_i) error();
STORE(NIID);
RPI = 1;
RPI = RPI * VI1;
```

a)

```
RPI = RPI * VI_j;
```

b)

Figure 3-7. Checking code added in the program

---

[1] A *control variable* of a program is a variable whose value can affect the CF of the program [43]. An *input control variable* of a loop-free interval is a control variable of the loop-free interval which is either already defined before the loop-free interval is entered or read as an input during the execution of the loop-free interval [43].

The second line in the code presented in Fig. 3-7 a) performs the check, which controls if the *NIID* corresponds to the currently traversed loop-free interval *ID* ID_i: if not, then an error is detected, otherwise, *CIID* variable is set to *NIID*. Than the input control variables of the loop-free interval are stored and the *RPI* is initialized to 1.

In each BB v_j entry the value of *RPI* is multiplied by the corresponding vertex identifier as it is presented in figure 3-7 b) (we will name vertex identifier of *BB* v_j - VI_j). In Fig. 3-7 ID_i and VI_j are constant values.

If a module call instruction is present in a BB, the *NIID* variable is updated to the first loop-free interval *ID* of the called module before the call instruction. After the module call instruction the check is performed if the *NIID* value corresponds to the last interval's *ID* of the called module. Before the module exit or the program stop the *TAB* function is performed to check the correctness of the last loop-free interval execution.

In order to reduce the memory and performance overhead of their method, the authors suggest to introduce an independent processor called *supervisory processor* for performing the CFC. The supervisory processor allows to separate the execution of the most time consuming CFC operations from the execution of the object program. It is recommended that the supervisory processor have much higher reliability than the processor executing the program, which allows avoiding the failure of the checking process itself.

In the presented CFC approach the introduction of the loop-free intervals gives two advantages: first, infinite paths are excluded, and second, the total number of paths is significantly reduced.

3.2 Experimental results

In order to evaluate the method memory and performance overhead some experiments were performed [43]: five Fortran applications were hardened according to the presented approach; the size of the applications varied from 23 to 103 source lines. The measured memory overhead varies from 90% to 175% for the considered applications and equals in average to 123.6%. For evaluation of performance overhead 6-7 runs of each application were performed; the performance overhead measured during these runs varied in average from 69.6% to 87%.

3.3 Advantages and limitations

The method is capable to detect CFEs of types 1, 2 and 3 (according to the fault model presented in section 1). It guarantees the detection of single CFEs of these types for erroneous branches inside the loop-free interval.

However, it seems that the method can miss erroneous inter loop-free interval branches (even if the probability that this event happens is quite low). It can happen if an erroneous jump leads from one loop-free interval ID_i to another loop-free interval ID_j so that:

- on the loop-free interval ID_j exit the *RPI* is equal to *RPI* expected on the exit of the loop-free interval ID_i in case of no error.
- the loop-free interval executed after the loop-free interval ID_j is the same as the loop-free interval which should be executed after loop-free interval ID_i in case of no error.

Assigning distinct vertex identifiers to all BBs in the program, not just in loop-free interval, can easily eliminate this problem.

As no checking code is added in the BB exit, the CFEs of type 4, which lead from some point of one BB to the beginning of a correct BB, are not detectable by the method. As the method is oriented to the inter-block CFEs detection it also does not detect errors of type 5.

The need to store the data-base leads to an additional memory overhead for the method. On the other hand the supervisory processor and loop-free intervals introduced in the method aim at memory and performance overhead reduction.

Other drawbacks of the method are:

- it does not seem easy to automatically implement the method, since besides CFG construction (needed by most of the CFC approaches) the method needs to partition the program into loop-free intervals and to build the data-base,
- the multiplication operation used to obtain *RPI* is rather time consuming,
- loop-free interval level detection introduces error detection latency.

4.   CFE DETECTION IN SEQUENTIAL AND PARALLEL PROGRAMS

4.1   The approach

In [48] a method is proposed, which aimed in particular at the detection of CFEs leading to wrong program module selection in a uniprocessor case or to an incorrect process to run selection in a multiprocessor case.

The method considered in [48] is particularly suited for structured programs, containing a large number of procedures dedicated to solve parts of the program task as well as for parallel programs. The method is intended to be applied to assembly-level programs.

A signature is associated to the main program as well as to each procedure, which is the symbolic name of the program/procedure. The program name is embedded into the program code and the procedure name into procedure code during the compilation time. Each name is represented in fixed length binary representation. A special register R is reserved to contain the run-time signature, *i.e.*, the name of the procedure under execution.

Code is added in the program during the compilation, which on each program/procedure entry and exit checks the run-time signature for consistency with the program/procedure name embedded in the code. In case of mismatch an error is signaled. In case of a long program/procedure it is suggested to perform more than one consistency check in order to reduce the error latency. How often the name is checked is thus a problem of trading-off between error detection latency and memory and performance overhead. Authors of the approach suggest that a check before each procedure call is a good choice.

On the program start the name of the program is put in the especially reserved register R. Before each procedure call some code is added, which moves the content of register R (*i.e.*, the currently executing procedure name) in a reserved place (*e.g.*, in a stack) and the name of the procedure to be executed next in register R. After the procedure is executed the name of the procedure is popped back to register R.

The proposed approach can be directly extended for identifying CFEs which lead to jumps over synchronization points in parallel programs.

4.2 Experimental results

In order to estimate the proposed approach experiments were performed in [48] on a 80386-based PC resorting to an in-house developed application containing 5 simple routines, which may call each other.

Fault injection was performed by means of a TSR (Terminal-Stay-Resident) program, which can be called by left-shift key during the program execution and which causes the processor to jump to a random location in the program memory space. During fault injection 300 faults were injected, among which 57% were determined by means of some detection mechanism embedded in the processor and 34% were detected by the proposed method. The estimated memory overhead in the considered case was 34%.

4.3 Advantages and limitations

The method proposed in [48] is transparent to the user and easy to implement, and it does not need a complete program CFG analysis. The user has the possibility to trade-off between redundancy, CFEs coverage and detection latency by choosing the number of checks to be performed inside the program/procedure. The main limitation of the method is its reduced CFEs coverage: in the monoprocessor case it is able to check only the correctness of the CF between program procedures.

5.   BEEC AND ECI

5.1   The approach

In this section the CFC approach described in [42], [49] is presented. This approach is composed of two independent CFC methods, which can be applied together: they are *Block Entry Exit Checking* (BEEC) mechanism [49] (this approach is based on the Block Signature Self Checking approach, presented in [42]), which checks the CF between program BBs and *Error Capturing Instruction* (ECI) mechanism, which inserts trap instructions in the data area and in the unused area of memory. In this way, if the program starts to fetch instructions from the unused or data memory areas, an error is detected. The idea of the ECI technique is based on that proposed in [50][51].

To increase the CFEs detection coverage it is suggested in [42], [49] to combine software BEEC and ECI techniques with a watchdog timer (WDT), as WDT is able to detect CFEs not detectable with software BEEC and ECI mechanisms (for example errors, which affect the CPU's capability to execute program code). In the following subsections 5.2 and 5.3 the BEEC and ECI techniques are presented, respectively.

5.2   BEEC

In this approach the program is partitioned into BBs and each BB is supplied with checking instructions as shown in Fig. 3-8.

At the beginning of each BB the call instruction to a routine (named *entry*) is added. At the end of each BB the call instruction to another routine (named *exit*) and embedded signature are added.

The entry routine checks if the execution of the previous BB was successfully completed by comparing the value stored in a static buffer with the unique *KEY* value. This *KEY* value is stored in a static buffer by the previous BB exit routine. If the check is successful the address of the first BB instruction ($m+1$) is stored in the static buffer; otherwise the CFE is detected.

A BB signature, which is equal to the sum of the size of the BB and the size of a call instruction, is stored after the exit routine call in the bottom of the BB. The exit routine sums the value ($m+1$) stored by the entry routine in the static buffer with the embedded BB signature ($n+k$), where n is the number of bytes in BB and k is the number of bytes of the exit routine call instruction, and compares the obtained value with the address of the last BB instruction ($m+n+k+1$). In the case of a successful comparison the exit routine stores the unique *KEY* in the static buffer and modifies the return address to the program in order to skip the embedded signature. In case of mismatch a CFE is detected and an error handling routine can be called from exit routine to initiate the recovery.

As it is possible to have different BBs with the same size, the BB signature computed as the size of the BB plus the size of a call instruction is not unique. However, the address of the BB's first instruction is unique for the program and consequently the value ($m+n+k+1$), which is compared with the BB embedded signature address is unique for each BB.

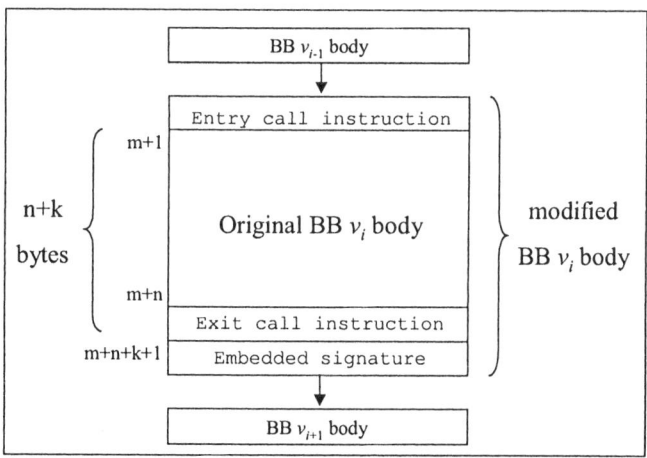

Figure 3-8. Checking instructions according to BEEC approach.

It is suggested to implement the technique by means of a postprocessor, which inserts BEEC instructions into the code generated by compiler.

To reduce the overhead caused by BEEC approach the authors of the method suggest to harden only the BBs with more than s instructions (in the experiments described in [49] s is equal to 5).

5.3   ECI

In this approach some special instructions (named ECIs) are stored in the memory locations not used during normal program execution. The execution of an ECI indicates a CFE. ECIs are inserted in the data area and unused area of memory. The ECIs can also be inserted in the program code area; in this case ECIs should be skipped over during the normal program execution.

Some instructions are proposed to implement ECIs:
- Software interrupt instructions.
- Unconditional branch instructions.
- Call instructions.
- Jump instructions.
- No-operation instruction (NOP).

These instructions can be used to initiate an error handling routing or to initiate an infinite loop; in the latter case a watchdog timer may be used for error detection. If the microprocessor has undefined operation-code detection in its design, then undefined operation codes in the microprocessor can also be used to implement ECIs.

5.4   Experimental results

In order to evaluate the proposed techniques some experiments were performed in [49] on a Motorola MC6809E microprocessor running 3 application programs.

The combination of BEEC, ECI and WDT was evaluated during experiments. In order to reduce the overhead only BBs, which contain more than 5 instructions were hardened with the BEEC technique.

The following overheads were measured: in the average the overhead in terms of program size was around 21.8% and the execution overhead around 99.6%. The ECI mechanism increased the data size by around 6.5% in the average.

Two fault injection methods were used for physical injection of transient faults: Heavy-Ion Radiation (HIR) and Power-Supply Disturbance (PSD).

During fault injection 6,000 errors were injected (1,000 for each of the 3 applications and for each of two fault injection methods). Some results of the performed fault injection campaigns are reported in Table 3-2. The table reports the average percentage of detected errors: the contribution of each of the techniques is indicated. The vector addresses of the interrupts unused during the experiments were provided with the address to a detection routine; this allowed to detect around 1.9% of errors. Besides this, the table reports the percentage of injected errors, which leaded to CFEs and the percentage of detected CFEs. Some undetected errors did not influence the result, so the correct outputs were produced; other undetected errors caused illegal RESET; the percentage of undetected errors producing wrong result is presented in the last column.

Table 3-2. Fault injection results (figures are approximated)

	Detected errors (%)					CFEs (%)	Detected CFEs (%)	Wrong result (%)
	BEEC	ECI	WDT	HW/SW Interrupts	Total			
HIR	42.7	11.5	23.7	3.4	81.4	88.8	89.5	9.3
PSD	43.2	8.8	41.0	0.4	93.3	96.9	96.0	3.3
Total	42.9	10.2	32.4	1.9	87.3	92.8	92.9	6.2

5.5   Advantages and limitations

The method is able to cover CFEs belonging to types 1-4 from the fault model presented in section 1 of this chapter; moreover, the ECI mechanism allows to cover erroneous branches, which have as their destination an unused area of memory. This method does not cover intra-block CFEs (errors of type 5). Unfortunately, in the papers presenting the BEEC mechanism it is not explicitly described how the exit routine assigns the *KEY* value of the next BB in case the current BB has more than one successor.

6.   EXPLOITING INSTRUCTION LEVEL PARALLELISM: ARC TECHNIQUE

6.1   The approach

In this section the technique named *Available Resource-driven Control flow monitoring* (ARC) [52] is described. ARC is a signature-monitoring technique applicable to assembly-level programs.

The particular feature of this technique is that it is oriented to processor architectures exploiting instruction level parallelism (ILP) in order to achieve higher performance. Particular focus in the described method is put on *Very Large Instruction Word* (VLIW) architectures; however, it can be adopted to other architectures exploiting ILP.

Processors with VLIW architecture contain multiple Functional Units (FUs), which allow performing more than one operation at a time. In the VLIW architecture the operation parallelism is identified statically by the compiler, which generates instructions composed of multiple operations that can be performed simultaneously on different FUs. Usually, VLIW processors have idle FUs during the program execution as either the program under execution does not possess the parallelism necessary to occupy all the FUs, or the compiler is not able to identify a sufficient number of operations, which can be executed in parallel, or for both reasons. In [52] the estimation was performed for the processor with VLIW architecture Multiflow TRACE 14/300, which showed that in the average the utilization of all resources is low (10-30% in the performed experiments). Even if this figure can be higher in different VLIW architectures, it is still expected that the percentage of idle resources during the program execution be rather high. The idea of the ARC approach is to get use of these idle resources in order to perform CFC with low performance overhead by scheduling checking code in idle resources.

In the ARC method program instructions are grouped in blocks and a block identifier *id* is assigned to each block. The ARC method operates with blocks, which are constructed based on the available idle resources.

An additional code is integrated in the program, which monitors the program CF. This code performs two tasks, which are

- *tracking task*, which updates the block signature during the program execution.
- *checking task*, which checks the run-time signature during the program execution.

A register named *key* is dedicated to contain the currently traversed program block *id*. In all the program entry points the *key* value is initialized with the 0 value; in the block boundary the *key* value is updated by tracking task to the subsequent block *id*.

Operations performing tracking and checking tasks are allocated in ARC in such a way that they use as much idle resources as possible.

The allocation of the operations required by the two tasks is performed as follows:

- *Checking task operations allocation:* the allocation of the checking operations is performed before the allocation of the operations performing the tracking task because the constraints for placement of checking operations are more restrictive. The checking operations are located in such a way that the time to reach the checking operations from any point of the program is bounded.

To reach this objective checking operations are allocated obeying the following constrains:

- A checking operation is added to each loop (the place of the checking task operations in the loop does not matter).
- A checking operation is added at each program exit point.

During the checking task operations allocation as much idle resources are used as possible. In the case the idle resources are not enough new instructions are added in the program.

- *Tracking task operations allocation.* The tracking task pursues the following objectives:
  - The program is partitioned in the smallest possible blocks while obeying the condition that idle resources are used, only. To reach this objective a block boundary is added each time the idle resources are available for allocating the tracking operations. In order to simplify the tracking task the following condition is fulfilled during the program partitioning to blocks: all successors of a program instruction belong to the same block (this condition simplifies the tracking task as the *id* of the current block should be always modified to the same *id* not depending on program state).

    Before the program partition to the blocks the *key* value is initialised to the 0 value in the program entry point: if the idle resources are not available the new instruction is added before the program entry point, which becomes the new program entry point and where the *key* initialisation is performed.
  - The mapping function *f*, which transforms the *key* from the old value to the new one, should be chosen in such a way that only the *id* of the block where the modification is performed is mapped into the *id* of the immediate successor block. To reach this objective the authors of the method suggest to use as *f* an injective function (*i.e.,* a one-to-one function), such as, for example, an integer add or subtract.

6.2   Experimental results

Experiments were performed [52] on the processor with VLIW architecture Multiflow TRACE 14/300 using 4 benchmark applications written in C and FORTRAN programming languages in order to estimate the method memory and performance overhead. Experiments show that for the considered benchmark programs 100% of the monitoring operations were scheduled in idle resources. The estimated performance overhead is thus close to 0 for all benchmarks and the memory overhead varies from 1.1% to 23.2%.

6.3   Advantages and limitations

The ARC approach allows to utilize idle resources available during the program execution on microprocessors with an architecture exploiting ILP for performing CFEs monitoring; this allows to keep performance overhead of the method quite low.

As the method does not operate with BBs but with blocks constructed according to different rules we will evaluate the ARC method with respect to the fault model presented in section 1, where the BB is changed with the block in the sense specified by the ARC method. With this assumption the ARC method is able to detect CFEs of types 1, 2, 4 and most of CFEs of type 3. It does not detect CFEs of type 3, which lead to erroneous branches having as their destination some instruction of BB body correct according to CFG. ARC method does not detect CFEs of type 5. As the ARC method block can include branch instructions the probability of intra-block CFE in blocks constructed according to the ARC method is higher with respect to intra-block CFEs in the case of program partition into BBs. However, the ARC method is able to detect some intra-BB CFEs if the source and destination of the erroneous branch belongs to different ARC blocks.

The ARC method introduces a non negligible error detection latency, as the *key* value checks are necessarily performed only at loops and at program exit points.

7.   VASC

7.1   The approach

In [53] a method of software signature-monitoring technique applicable at the assembly-level and named *Versatile Assigned Signature Checking* (VASC) is presented. The method has been proposed for both mono- and multi-processor system, although the reported experimental results only cover the latter case.

The main particularity of the VASC method consists in the definition of the program logical blocks. The aim of this definition is to have blocks of the desirable size; the method proposes to vary also the size of the checking intervals.

In the VASC method each block may consist of an arbitrary number of sequential (in dynamic sense) instructions. The block can include branch instructions, program procedure call and return instructions, and so on. In order to keep the tracking task simple all successors of each program instruction are always included in the same block (like in the ARC method).

To each program block a signature *blockID* is assigned. In the block boundaries instructions are added, which update the run-time signature from the current blockID to the blockID of subsequent block according to the program CF. The consistency of the run-time signature and the assigned signature of the currently traversed block are periodically controlled. In the case of discrepancy an error is detected.

The number of instructions in the block is limited by the *block size* value and number of instructions between two check operations is defined by the *check interval* value. The block size as well as the check interval are defined by the user. These two values are completely independent, which means that the block size can be greater or equal to the check interval, and vice versa. In order to obtain the flexibility for improving the placement of checking and tracing operations, some tolerance can be specified by the user for the block size as well as the check interval values. As an example, if the block size is defined as 7 instructions with a tolerance of 2 instructions, then the block size can vary from 5 to 9 instructions.

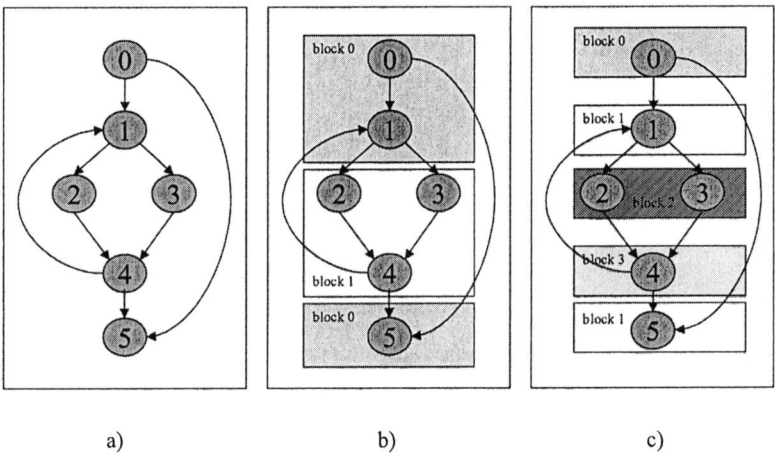

a)   b)   c)

Figure 3-9. Example of program partition to blocks according to the VASC method

In Fig. 3-9 an example of program partition to blocks according to the VASC method is presented. In Fig. 3-9 a) the program graph is presented (here nodes correspond to program instructions). In Fig. 3-9 b) the program partition to blocks for block size equal to 3 is presented. In Fig. 3-9 c) the program partitioning to blocks for block size equal to 2 is presented: in this case blocks can not be further enlarged. For example, block 0 can not be enlarged because its node 0 has two successors (nodes 1 and 5), which should belong to the same block; subsequently they cannot be included in the block 0 because otherwise its size would overcome the size 2.

7.2 Experimental results

In order to evaluate the proposed technique experiments were performed on two systems based on a Transputer T805 processor and a PowerPC processor [53]. Both of them are distributed memory parallel systems with four nodes.

The performance and memory overhead of the method were measured on both processors resulting to three applications. Experiments consider different combinations of block size and check interval. The authors of the method considered block size smaller than the check interval as checking operations involve a higher overhead with respect to the tracking operations and, besides, usually CFEs are propagated by tracking operations and can be detected in the following, when checking is performed.

In the experiments block size and check interval varied from block size equal to 30 and check interval equal to 50 to block size equal to 100 and check interval equal to 120; for the considered applications the code overhead varied from 2% to 0.5% in the average, and the execution overhead varied from 14% to 2.5% in the average for the T805 processor; for the PowerPC processor the code overhead varied from 6.5% to 2.5% in the average, and the execution overhead varied from 11% to 3.5% in the average. As expected, in general both the performance and the memory overhead decrease when the block size and the check interval increase.

The code size overhead is smaller for the T805 processor than for the PowerPC processor as the signature-monitoring instructions in the T805 processor are small, while in the RISC processor all instructions have fixed size; however, the performance overhead for the T805 processor is higher than for the PowerPC processor, as the monitoring operations operate with registers in the PowerPC processor and with a slower cache in the T805 processor.

Fault coverage figures were measured only for the PowerPC processor on three applications by using fault injection. Fault injection was performed using the Xception fault injection tool [54], which is able to emulate hardware transient faults in the processor FUs (Data Bus, Address Bus, Floating Point Unit, Integer Unit, General Purpose Register, Branch Processing Unit, Memory Management Unit) and in the memory.

Applications were hardened with the VASC technique with block size equal to 10 and check interval equal to 18. Only the source code was hardened, while the library code remained unhardened; hardening the library code as well can thus further increase the FC figures (and the overhead).

During the experiments 10,000 transient faults have been injected; each of the faults was injected in one randomly selected processor of the system and caused one or two bit flips of a randomly chosen processor FU. Fault injection experiments performed on 3 applications hardened with the VASC mechanism showed that the built-in processor fault detection mechanism was able to detect in the average 37.6% of the injected faults and the VASC mechanism was able to detect in the average 4.7% of injected faults; in the average 8.3% of the injected faults remained undetected. The overall increase of fault detection achieved by hardening the applications with the VASC technique was in the average equal to 3.2%. Study of undetected faults showed that these faults are mainly pure data faults; to detect these faults some data detection mechanism (a variety of them is presented in chapter 2) should be combined with VASC. Authors of [53] note that the percentage of undetected faults is high also because experiments were performed on RISC processors, where faults cause more data errors and less CFEs (while in CISC processor the situation is the opposite).

The VASC fault coverage for different block size and check interval was also studied in [53]. The performed experiments showed that the fault coverage is not strictly decreasing with the growing block size and check interval. Authors of [53] give some responsibility for such behavior to interaction of VASC with the built-in fault detection mechanism.

Study of the experimental results showed that for obtaining a good fault coverage the check interval should be less than 50 instructions and the block size should be kept small.

7.3   Advantages and limitations

The VASC method gives to a user the possibility to vary the block size and the check interval and consequently to trade-off the method detection capabilities and overhead.

When BBs are substituted with blocks in the sense of the VASC method, the VASC method is able to detect CFEs of the same types as the ARC method (see section 6.3).

The probability of intra-block CFE, in blocks constructed according to VASC method is higher with respect to intra-block CFE, in case of program division to BB. However, the VASC method is able to detect some intra-BB CFEs if source and destination of an erroneous branch belong to different blocks (the reasons are the same as for the ARC method). The VASC method (as well as the ARC method) introduces some error detection latency.

8.   ECCA

8.1   The approach

The software-implemented CFC approach named *Control flow Checking using Assertions* (CCA) and its enhanced version (ECCA) are presented in

[45], [55], [56]. In this book we describe only the ECCA approach as it has improved characteristics with respect to CCA. ECCA has two versions: one version is oriented to programs coded in high-level language, and another one is oriented to intermediate-level representations. We will denote the two versions as ECCA-HL and ECCA-IL, respectively.

In order to reduce the overhead, ECCA divides the program into a set of blocks, where the block is a collection of consecutive BBs (BBs are called *Branch Free Interval* or BFI in [45], [55], [56], but we will hold on accepted terminology) with single entry and single exit. The shorter the block is, the higher the fault coverage is and the lower the error detection latency is, whereas the memory and performance overhead is higher. By properly choosing the block length it is possible to find the most suitable trade-off for the user purposes.

8.2 ECCA-HL

Error detection in ECCA-HL is performed reasoning to the exception handler.

ECCA-HL assigns a unique prime number identifier (called *Block Identifier* or BID) greater than 2 to each block of a program. During program execution the global integer variable *id* is updated to contain the currently traversed block identifier.

Two assertions are added to each block:

- a *SET* assertion is added at the beginning of the block, which executes two tasks: it assigns the *BID* of the current block to the *id* variable and it checks if the block the execution came from is a predecessor block, according to the CFG. A divide by zero error signals a CFE. The SET assertion implements the following formula:

$$id = \frac{BID}{\overline{(id \bmod BID)} \cdot (id \bmod 2)}, \tag{1}$$

$$\text{where } \overline{(id \bmod BID)} = \begin{cases} 1, & \text{if } (id \bmod BID) = 0 \\ 0, & \text{if } (id \bmod BID) \neq 0 \end{cases}$$

- A *TEST* assignment is executed at the end of the block and executes two tasks: it updates the *id* variable taking into account the whole set of successor according to CFG blocks and checks if the current value of the *id* variable is equal to *BID*. The TEST assertion implements the following formula:

$$id = NEXT + \overline{(id - BID)} . \tag{2}$$

The variable *NEXT* is equal to the product of BIDs of all successors according to the CFG blocks of the current block, *i.e.*,

$$NEXT = \prod BID_{successor} . \tag{3}$$

$$\overline{(id - BID)} = \begin{cases} 1, \text{if } (id - BID) \neq 0 \\ 0, \text{if } (id - BID) = 0 \end{cases} .$$

The *NEXT* and *BID* variables are generated once before the program execution, whereas the *id* variable is updated during the program execution.

As an example, in Fig. 3-10 the program code from Fig. 3-1 a), hardened according to ECCA-HL, is reported.

```
id = BID0;
i = 0;
id = BID1*BID5 + !!(id-BID0);
while(i < n)
{
  id = BID1/((!(id%BID1))*(id%2));
  id = BID2*BID3 + !!(id-BID1);
  if (a[i] < b[i])
  {
    id = BID2/((!(id%BID2))*(id%2));
    x[i] = a[i];
    id = BID4+!!(id-BID2);
  }
  else
  {
    id = BID3/((!(id%BID3))*(id%2));
    x[i] = b[i];
    id = BID4+!!(id-BID3);
  }
  id = BID4/((!(id%BID4))*(id%2));
  i++;
  id = BID1*BID5+!!(id-BID4);
}
id = BID5/((!(id%BID5))*(id%2));
```

Figure 3-10. Example of application of ECCA-HL

8.3   ECCA-IL

In order to fulfill the language portability requirements the authors of ECCA also proposed ECCA-IL. This technique works at the RTL stage used by the GNU's compiler as intermediate-level representation. ECCA-IL takes advantage of the property that a block in RTL can have two successors at most. Therefore, the method rewrites the SET and TEST assertions using a cheaper (in terms of CPU time) variant.

Each SET assertion in ECCA-IL implements the following formula:

$$r_1 = (r_1 - BID) \cdot (r_2 - BID), \tag{4}$$

$$r_1 = \frac{BID + 1}{\left(\frac{r_1 + 1}{r_1 \cdot 2 + 1}\right)}, \tag{5}$$

where r_1 and r_2 are global registers. Under correct program execution one of the registers r_1 and r_2 contains the *BID* value. After the first statement is executed the r_1 register takes the value 0 in the error free case and not null in the case of error. After execution of the second statement the r_1 register takes the value BID +1 in the case of no error. In case of error, r_1 is different than 0, and as a result the division $(r_1+1)/(r_1\cdot2+1)$ among integer values will provide a result equal to 0. As a consequence, a divide-by-zero error will be originated, signaling the presence of a CFE.

The version of the SET assertion to be used in the intermediate-level is presented in Fig. 3-11.

```
tmp1  = BID + 1
tmp2  = r1 - BID
tmp3  = r2 - BID
r2    = tmp2 × tmp3
tmp2  = r2 + 1
tmp3  = r2 << 1
tmp3  = tmp3 + 1
tmp2  = tmp2/tmp3
r1    = tmp1/tmp2
```

Figure 3-11. Intermediate-level version of the SET assertion

If the program is fault-free, the execution time of the SET assertion is relatively small due to the fact that multiplication is by zero and both divisions are by 1.

The TEST assertion in the intermediate-level corresponds to the following two assertions:

$$r_1 = (r_1 - BID) \cdot NEXT_1,$$
$$r_2 = (r_1 - BID) \cdot NEXT_2,$$

where $NEXT_1$ and $NEXT_2$ represent the BIDs of two successor blocks.

In the case of correct execution $(r_1 - BID)$ is equal to 1, and therefore registers r_1 and r_2 are set to $NEXT_1$ and $NEXT_2$, respectively; otherwise, r_1 and r_2 will be set to nonprime values different from the BIDs of the successor blocks of the current block and the CFE will be detected by the next executed block SET assertion.

Fig. 3-12 reports the intermediate representation of the TEST assertion.

```
tmp₁ = r₁ - BID
r₁ = tmp₁ × NEXT₁
r₂ = tmp₁ × NEXT₂
```

Figure 3-12. Intermediate-level version of the TEST assertion

If the CF is correct both multiplications of the TEST assertion are by 1.

8.4   Experimental results

In order to evaluate the proposed approach some experiments were performed in [45] using the FERRARI software-based fault injection tool [57] on a SUN SPARC workstation.

During the experiments single bit-flips were randomly injected in registers and in code memory (including libraries). In all the experiments over 400,000 errors were injected; 3 applications were considered. The first two columns of Table 3-3 describe the transient error model considered in the experiments. The last column presents the approximate average percentage of undetected errors in the applications hardened with the proposed technique. Several mechanisms, including System detection, Timeout, User detection and the ECCA approach contributed to the detection of the injected errors.

The authors do not report figures showing the global percentage of undetected faults.

Table 3-3. Transient error model

Model name	Model description	Average rate of undetected errors (%)
AddIF	address line error resulting in executing a different instruction	*2.9*
AddIF2	address line error resulting in executing two instructions	*1.6*
AddOF	address line error when a data operand is fetched	*2.4*
AddOS	address line error when an operand is stored	*4.1*
DataIF	data line error when an opcode is fetched	*2.4*
DataOF	data line error when an operand is loaded	*7.6*
DataOS	data line error when an operand is stored	*4.4*
CndCR	errors in condition code flags	*7.2*

8.5   Advantages and limitations

The main merit of the approach is its high CF coverage. ECCA covers all single CFEs of types 1, 3 and 4 from the fault model presented in section 1. Legal but wrong branches (errors of type 2) as well as intra-block CFEs (errors of type 5) are not considered by the method.

The drawback of the method is the quite high memory and performance overhead: although only two instructions for block are added in ECCA-HL, this instructions are rather complex and are translated in a high number of instructions in the executable assembly code. Getting use of the special properties of the intermediate-level representation for simplifying the SET and TEST assertions in ECCA-IL help to overcome this problem.

9.   PLAIN INTER-BLOCK ERRORS DETECTION

9.1   The approach

In [58], [59] a software-implemented method for inter-block CF hardening is presented. This method is not aimed at complete CFEs coverage; rather it proposes an economical and easy to automatically implement approach for detecting a part of the CFEs. It is aimed to be combined with the instruction duplication approach for data hardening presented in chapter 2.

The method is based on the following rules applicable to programs coded in high-level language:

- An integer signature k_i is associated with every BB v_i in the program.
- A *global execution check flag* (*gef*) variable is introduced in the program for storing a run-time signature; an instruction, which assigns to the

variable *gef* the value k_i, is added at the beginning of each BB v_i; an instruction, which performs a consistency check between the variable *gef* value and the value k_i is added at the end of the BB v_i; in the case of mismatch an error is signaled.

- For each program condition instruction the test is repeated at the beginning of both BBs corresponding to the true and (possible) false clause. If the newly introduced test does not provide a positive answer, an error is signaled.
- An integer value k_j is associated with any procedure j of the program.
- Immediately before each *return* instruction of the procedure, the value k_j is assigned to the variable *gef*; a consistency check between the value of the variable *gef* and the value k_j is performed after any procedure call; a mismatch signals a CFE.

Fig. 3-13 presents the program code from Fig. 3-1 a), hardened according to the technique considered in this section.

```
gef = 0;
i = 0;
if (gef != 0) error();
while(i < n)
{
    if (i >= n) error();
    if (a[i] < b[i])
    {       if (a[i] >= b[i]) error();
        gef = 1;
        x[i] = a[i];
        if (gef != 1) error();
    }
    else
    {       if (a[i] < b[i]) error();
        gef = 2;
        x[i] = b[i];
        if (gef != 2) error();
    }
    gef = 3;
    i++;
    if (gef != 3) error();
}
if(i < n) error();
```

Figure 3-13. The technique application

9.2   Experimental results

In order to evaluate the effectiveness of the proposed technique experiments were performed and reported in [59].

Experiments were performed on a T225 transputer and on three C programs. Applications were hardened with the technique described in this section as well as with the instruction duplication technique described in [58], [59] and presented in chapter 2. The measured code size overhead was around 4 times the original code and the performance overhead was ranging from 2.1 to 2.5 times.

Software fault injection campaigns were performed in order to evaluate the robustness of the hardened programs. Bit-flips were separately injected in the memory area containing the program code and in the memory area containing the program data. In each experiment 1,000 faults were injected in the original program; in the modified program 1,000 faults multiplied by the memory size increase factor were injected. Experiments showed that around 56.7% in the average of the injected faults were detected by the proposed software approach in code memory and around 52.8% in average in the data memory; in the code memory percentage of undetected faults leading to wrong answer reduced from around 45.5% in average in unhardened programs to around 0.2% in the average in the hardened programs, and in the data memory from around 77.5% in the average to 0% in average; percentage of time-out reduced from around 10.5% in average to around 1.1% in average in the code memory.

To get more confidence radiation experiments were also performed. During these experiments only the program memory was exposed to faults. One application was considered. Experiments showed that for the hardened program around 2.1% of wrong answers were produced and around 0.6% of time-out.

9.3   Advantages and limitations

The main advantage of the proposed method of CFEs detection is its simplicity. The main drawback is the incomplete CFEs coverage. The method is not able to cover erroneous branches having as their destination the first BBs' instructions.

The method is able to cover the CFEs of types 2, 3 and 4 from the fault model presented in section 1. Only those CFEs of type 4 can be detected, which cause branches skipping checking code in the top of the destination BB. This method does not detect CFEs of type 1 and intra-block erroneous branches (CFEs of type 5).

10. CFC VIA REGULAR EXPRESSIONS RESORTING TO IPC

10.1 The approach

In [61] a signature-monitoring approach is proposed, where CFC is implemented by exploiting the characteristics of a multiprocess/multithred operating system.

In this approach a unique *block symbol* is assigned to each BB. Then each path of the CFG can be represented by a string of symbols, obtained as concatenation of block symbols corresponding to BBs included in the path. All block symbols form an alphabet A. All strings of symbols corresponding to legal paths (according to the program CFG) form a language $L = (A, R)^1$, where R is a regular expression, able to generate these strings of symbols.

Legal paths of the CFG represent correct CF executions.

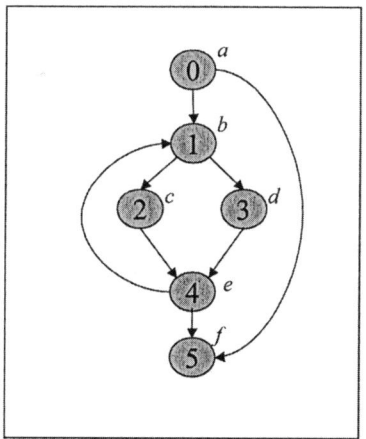

Figure 3-14. CFG with block symbols assigned to each BB

An example of a CFG with block symbols assigned to BBs is presented in Fig. 3-14. For this example $A = (a, b, c, d, e, f)$ and $R = a(b(c|d)e)^*f$.

In this example, if the program execution produces the string $S =$ "*abdef*", then this execution belongs to language $L = (A, R)$ and it is correct, whereas the string $S =$ "*abcdef*" does not belong to language $L = (A, R)$, and consequently the CF has an error.

---

[1] For more details on language L and regular expression R see [60].

The checking process uses multiprocess/multithred programming facilities provided by operating systems. The referenced program and the checking program are defined as two different processes, which communicate using Inter Process Communication (IPC) facilities. The check program controls if the input string belongs to the language *L*. During program execution a string composed of symbols of BBs being traversed is generated by the referenced program and is transmitted to a check process using the IPC. A check process controls if the string received from the main program belongs to the language *L* and detects a CFE if it is not. For generation of block symbols some suitable instructions are added in the end of each BB in the referenced program.

The proposed method can be applied to programs described on high-level language or assembly-level language.

The method permits to trade off between error detection latency and performance and memory overhead: only one check (after the program execution) could be performed, if the occurrence of an error during the program execution is not critical. On the other hand, if early error detection is wanted, each BB can be split to several sub-blocks.

10.2   Experimental results

In order to evaluate the proposed approach some experiments were performed, where the Windows 2000 operating system and ad-hoc fault injector were used. Experiments were performed on 5 applications. During experiments transient errors were injected in the code segment of the applications. In the experiments a check was performed at each BB.

The following figures were obtained (all figures correspond to the average over 5 applications): memory overhead around 114.3%, time overhead around 172.4%, around 3.2% of all injected errors were detected by the proposed approach, while around 89.0% of injected errors resulted in application crash.

Most of the other undetected errors leaded to a crash of the applications (around 89.0%).

10.3   Advantages and limitations

The proposed approach is able to detect all CFEs of type 1 and some CFEs of types 3 and 4. It does not detect those CFEs of type 3 which cause erroneous branches leading from one BB to any point of one of its successor BB body. It does not detect those CFEs of type 4 which cause erroneous branches, which source and destination BBs are both successors of the BB that is the predecessor of the erroneous branch source BB the execution came from. The example of such CFE is presented in Fig. 3-15. If the execution came to the BB v_3 from the BB v_1 and the erroneous branch outgoes from the BB v_3 body, then such erroneous branch is undetectable by the approach.

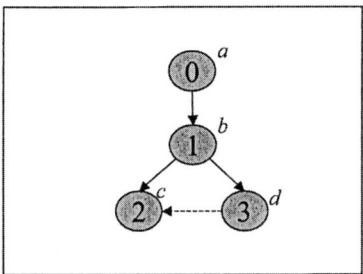

Figure 3-15. Example of CFE undetectable by the approach

This approach does not detect CFEs of types 2 and 5.

The main advantage of the approach is that it demands very low performance and memory overhead exploiting the multiprocess capabilities offered by the operating system; however, these capabilities are not always available.

11. CFCSS

11.1 The approach

In [44] an assembly-level CFC approach named *Control Flow Checking by Software Signatures* (CFCSS) is proposed.

CFCSS assigns a unique arbitrary number (signature) s_i to each BB. During program execution a run-time signature G is computed in each BB and compared with the assigned signature. In the case of discrepancy a CFE is detected. A run-time signature G is stored in one of the general-purpose registers (GSR).

At the beginning of the program, G is initialized with the signature of the first block of the program. When a branch is taken the signature G is updated in the destination BB v_i using the signature function f. The signature function f is computed resorting to the following formula:

$$f(G, d_i) = G \oplus d_i, \tag{6}$$

where

$$d_i = s_j \oplus s_i, \tag{7}$$

and s_j is the signature of the predecessor of the BB v_i. d_i is calculated in advance during the compilation and stored in the BB v_i.

For example, if the currently traversed BB is s_j, then $G = s_j$; when the control is passed to the BB s_i, G is updated as follows:

$$G = G \oplus d_i,$$

substituting the values of G and d_i we have

$$G = s_j \oplus s_j \oplus s_i = s_i.$$

Therefore, in the absence of CFEs the variable G contains the signature of the currently traversed BB.

If a CFE happened, leading to an illegal branch from BB v_k to v_i, (whose predecessor is the BB v_j) then $G = G \oplus d_i = s_k \oplus s_j \oplus s_i \neq s_i$.

At the top of each BB v_i (before the original instructions of the BB) some new code is added, which updates the signature G using the signature function f and compares the computed run-time signature with the assigned one (Fig. 3-16). In the case of mismatch the error is detected and the control is transferred to an error handling routine.

```
G = G ⊕ d_i;
if(G != s_i) error();
```

Figure 3-16. Checking code

If the BB v_i has more than one predecessor (*i.e.*, v_i is a branch-fan-in BB) an adjusting signature D is defined in each predecessor BB of v_i and used in the BB v_i to compute the signature. The adjusting signature D is set to 0 for one arbitrary chosen predecessor BB of v_i (let it be v_j); for each BB v_k, $k \neq j$ the predecessor of v_i, the adjusting signature D is defined as $D = s_j \oplus s_k$. For the BB v_k (predecessor of the branch-fan-in BB) the checking code is presented in Fig. 3-17. For the branch-fan-in BB v_i the checking code is presented in Fig. 3-18.

```
G = G ⊕ dₖ;
if (G != sₖ) error();
D = sⱼ ⊕ sₖ;
```

Figure 3-17. Checking code for predecessor BB of the branch-fan-in BB

```
G = G ⊕ dᵢ;
G = G ⊕ D;
if (G != sᵢ) error();
```

Figure 3-18. Checking code for the branch-fan-in BB

As an example, in Fig. 3-19 the program code from Fig. 3-1 a) modified according to CFCSS technique, is reported.

```
G = s0;
if (G != s0) error();
D = 0;
i = 0;
while(i < n) {
  G = G ^ d1; G = G ^ D;
  if (G != s1) error();
  if (a[i] < b[i])
  { G = G ^ d2;
    if (G != s2) error();
    D = 0;
    x[i] = a[i];
  }
  else
  { G = G ^ d3;
    if (G != s3) error();
    D = s2 ^ s3;
    x[i] = b[i];
  }
  G = G ^ d4; G = G ^ D;
  if (G != s4) error();
  D = s0 ^ s4;
  i++;
}
G = G ^ d5; G = G ^ D;
if (G != s5) error();
```

Figure 3-19. Program hardened with the CFCSS approach

Signatures are embedded in the program during compilation or preprocessing.

Once the CFE corrupted the CF causing the discrepancy between runtime signature and the expected one in some program BBs, the run-time signature remains different than the expected signature also in subsequent BBs. Basing on this property the authors of the CFCSS technique propose to perform consistency checks only in some of the program BBs, which allows to reduce the technique overhead. Postponing the check is possible only in case the error detection latency is acceptable for the application.

11.2   Experimental results

In order to evaluate the CFCSS technique experiments were performed and described in [44]. In the experiments 7 applications were considered and faults of 3 types were injected:

- branch deletion: a branch instruction is replaced with NOP instruction,
- branch creation: an unconditional branch is randomly inserted into the program,
- branch operand change: the immediate field of an instruction is corrupted.

Experiments were performed on a R4400 MIPS processor; 500 faults were injected. Experiments showed that the application of the CFCSS technique allowed to decrease the rate of incorrect undetected outputs from around 33.7% in average to around 3.1% in average. In the considered experiments the CFCSS technique introduced in the average around 45.1% of memory overhead and in the average around 43.1% of performance overhead.

11.3   Advantages and limitations

The proposed approach offers an economical way for CFEs coverage: an XOR operation used for run-time signature computation is less time consuming than multiplication or division.

The approach is capable to detect most CFEs of types 1, 3 and 4 according the fault model presented in section 1. It does not detect CFEs of types 2 and 5 and those CFEs of type 4 which lead from inside some BB to the beginning of some of its successor BB.

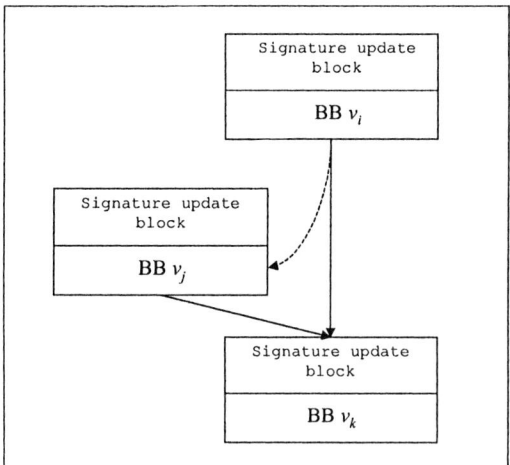

Figure 3-20. Example of CFE undetectable by CFCSS approach

Some CFEs of types 3 and 4 escape error detection. For example, let us consider the situation presented in Fig. 3-20. Here three BBs are presented (v_i, v_j and v_k). If a CFE of type 3 (or 4) happens, which introduces an erroneous branch (presented with dotted line in figure) leading from the end of the BB v_i (or from some instruction of the original BB v_i body in case of type 4 CFE) to some point in the BB v_j after the initial signature update block, then this CFE escapes detection by the CFCSS approach.

As it is shown in [44] aliasing errors is also possible for CFEs of type 1 in the case multiple BBs share multiple BBs as their destination nodes. For example, given a program CFG with the set of edges B containing the subset $\{b_{1,4}, b_{1,5}, b_{2,5}, b_{2,6}, b_{3,5}, b_{3,6}\}$, an erroneous illegal branch $b_{1,6}$ is not detectable by the method.

12. ACFC

12.1 The approach

In the work [62] a software-based signature-monitoring technique named *Assertions for Control Flow Checking* (ACFC) is presented.

In this method a bit of a special variable is associated with each program BB. This variable is named *execution status* (*ES*). In case of a big program it might be necessary to introduce more than one *ES* variables. Some additional code is added to the program, which sets the bit of the *ES* variable to the value 1 when the corresponding BB is traversed. When the program ends a check is performed, which controls the run-time CF correctness by comparing the *ES* variable value with a constant, whose value is 1 in all bits corresponding to the BBs, which should be traversed in the fault-free case.

In the beginning of the program the *ES* variable is set to 0. To set the bit of the *ES* variable to value 1 the XOR operation is used. In this way if the CFE causes the BB re-execution the corresponding bit is reset to the 0 value and the error can be detected during the check operation.

Some language constructs on the example of the C language are considered. In the case of *if-then-else* construct, where each branch contains only one BB, the same bit is associated to the BBs of the two branches, as only one of them should be executed in the fault-free case. In the case of the nested *if-then-else* construct and *switch* construct with *break* statement and *default* section (if *default* section is absent the dummy *default* section is added) the following solution is proposed. A bit of the *ES* variable is assigned to each entry BB and each exit BB of each construct branch. In the entry BB of the branch the bit corresponding to this BB and the bits corresponding to the exit BBs of the other branches are set to value 1. Similarly, in the exit BB of the branch the bit corresponding to this BB and the bits corresponding to the entry BBs of the other branches are set to value 1. In this way, if the CFE introduces a branch so that the entry BB v_i of one branch and the exit BB v_j of another branch are executed, then the error is detected, as the bits corresponding to the BBs v_i and v_j are set to 0. In order to detect CFEs inside the intermediate BBs of some branch, which are situated between the entry and exit BBs of this branch, an ES_k variable is introduced, whose value is checked before the branch exit BB. This technique is also extended to the *switch* construct, without *break* statement.

In case of the loop construct the check operation is performed in the last BB of the loop construct, so that the re-execution of the BB does not cause bits of the *ES* variable to be set to value 0. After the check operation the *ES* variable is set to the value it had before entering the loop. If a *break* or *continue* statements are used, then the check is performed before these statements and the *ES* variable is set to the value the variable should have in destination BB.

An example of application of the ACFC method to the program from Fig. 3-1 a) is presented in Fig. 3-21.

```
ES_1 = 0;
ES_1 = ES_1^01;
i = 0;
while(i < n)
{
  ES_1 = ES_1^10;
  if (a[i] < b[i])
      {
                    ES_1 = ES_1^100;
                    x[i] = a[i];
      }
  else
      {
                    ES_1 = ES_1^100;
                    x[i] = b[i];
      }
  ES_1 = ES_1^1000;
  if(ES_1 != 01111) error();
  ES_1 = 01;
  i++;
}
if(ES_1 != 01) error();
```

Figure 3-21. ACFC technique application example

12.2   Experimental results

In order to evaluate the proposed technique some experiments were performed by means of the in-house developed software-based fault injection tool SFIG: the gathered results were reported and analyzed in [62]. The SFIG tool is able to inject transient faults of the types presented in FERRARI [57] (see Table 3-3, columns 1 and 2). Experiments were performed on 5 application programs. During the experiments the proposed technique was compared with previously developed ones, namely, the ECCA and CFCSS techniques. The hardened according to the considered techniques versions of the programs were compiled with and without compiler optimization.

Table 3-4 presents average and rounded off overheads measured during the experiments for the considered techniques.

In all the experiments about 100,000 faults were injected. During the experiments faults are detected by following four mechanisms: operating system, time-out, user checks (*i.e.*, programmer-inserted debugging checks) and CFC technique. Experiments showed that the ACFC technique improved faults coverage by around 6% compared with unhardened programs. Around 87% in average of the faults were detected by the ACFC technique, which is around 1.5% in average less than faults coverage of CFCSS technique and around 4% in average less than faults coverage of ECCA technique (faults coverage figures are approximate as they are taken from graphical representation presented in [62]).

On the other side, experiments showed that the ACFC technique is less time and memory consuming with respect to ECCA and CFCSS techniques.

Table 3-4. Average memory and performance overheads comparison

	Memory overhead (%)			Performance overhead (%)		
	ECCA	CFCSS	ACFC	ECCA	CFCSS	ACFC
With compiler optimization	400.1	99.4	69.6	469.1	134.8	87.6
Without compiler optimization	176.3	48.2	34.4	154.2	43.7	30.5

12.3   Advantages and limitations

The ACFC method proposes an economical solution (in the sense of memory and execution overhead) to the CFEs detection problem. However, its CFEs coverage capabilities are limited. It only partially covers the CFEs of types 1, 3 and 4 and it does not cover the CFEs of types 2 and 5.

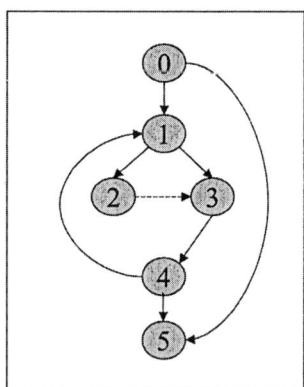

Figure 3-22. Example of CFE not detectable by the ACFC technique

Let us consider some examples of CFEs of types 1, 3 and 4 which are not covered by the ACFC technique. An example of CFE of type 1 which is not detectable by the method can be as follows: an illegal branch in the nested *if-then-else* construct, which leads from the end of the entry BB of some branch directly to the exit BB of the same branch skipping intermediate BBs.

An example of CFE of type 3 or 4 not detectable by the method is presented with a dotted arrow in Fig. 3-22. In this example it is supposed that checking code is executed in BB v_2 but skipped in BB v_3.

13.   YACCA

13.1   The approach

In [46] and [47] a software-implemented inter-block CF monitoring technique applicable to high-level program description and named *Yet Another Control flow Checking Approach* (YACCA) is presented.

The YACCA approach assigns to each program BB v_i two unique identifiers $I1_i$ and $I2_i$. The identifier $I1_i$ is associated to the BB v_i entry and the identifier $I2_i$ is assigned to the BB v_i exit.

An integer variable *code* is introduced in the program, which stores a run-time CF signature during program execution. The *code* variable is updated by means of the *set* assertion to the value of the entry identifier $I1_i$ at the beginning of the BB v_i, and to the value of the exit identifier $I2_i$ at the end of the BB v_i.

Before each set assertion a *test* assertion is performed. At the beginning of the BB v_i a test assertion verifies if the run-time value of the *code* variable corresponds to the identifier $I2_j$ of some BB belonging to the *pred*(v_i) set, while at the end of the BB v_i a test assertion verifies if the run-time value of the *code* variable corresponds to $I1_i$.

The update of the variable *code* value is performed according to the following formula:

$$code = (code \ \& \ M1) \oplus M2, \tag{8}$$

where $M1$ represents a constant mask whose value depends on the set of possible predecessor values of the *code* variable, whereas $M2$ represents a constant mask depending both on the identifier which should be assigned to the *code* variable and on the possible predecessor values of the *code* variable. For example, the values $M1$ and $M2$ can be defined as follows:

- for the *set* assertion at the beginning of the generic BB v_i:

$$M1 = \overline{\left(\underset{j:v_j \in pred(v_i)}{\&} I2_j \right) \oplus \left(\underset{j:v_j \in pred(v_i)}{\vee} I2_j \right)}, \tag{9}$$

$$M2 = (I2_j \& M1) \oplus I1_i. \tag{10}$$

- for the set assertion at the end of the generic BB v_i:

$$M1 = 1,$$

$$M2 = I1_i \oplus I2_i. \tag{11}$$

The binary representation of $M1$ obtained by Eq. (9) contains the value 1 in the bits having the same values in all identifiers $I2_j$ of BBs from $pred(v_i)$, and the value 0 in the bits having different values in these identifiers. The operation ($code$ & $M1$) allows to set the $code$ variable to the same value I from any possible predecessor value of the $code$ variable. Therefore, performing the XOR operation of I and $M2$ allows to obtain the value $I1_i$.

To avoid the aliasing effect the identifiers of the BBs should be chosen in such a way that the new value of the $code$ variable is equal to the targeted value if and only if the old value of the $code$ variable is possible according to the program CFG, *i.e.*, the operation ($I2_j$ & $M1$) should not return the value I if BB v_j does not belong to $pred(v_i)$.

The test assertion introduced at the beginning of the BB v_i with $pred(v_i) = \{v_{j1}, v_{j2}, ..., v_{jn}\}$ is implemented as follows:

```
ERR_CODE |= ((code != I2j1) &&

&& (code != I2j2) && (...) && (code != I2jn)),   (12)
```

where the *ERR_CODE* variable is a special program variable containing the value 1 if the CFE is detected and 0 otherwise. The *ERR_CODE* variable is initialized with the 0 value at the very beginning of the program execution.

The *test* assertion introduced at the end of the BB v_i is implemented as follows:

```
ERR_CODE |= (code != I1i).                       (13)
```

In order to identify the wrong branches the test is repeated for each conditional branch at the beginning of both the true and false clause. In order to identify all wrong branches each condition should contain the "else" clause; if the "else" clause is absent it should be introduced and the corresponding BB should contain *test* and *set* assertions.

Fig. 3-23 presents the result of the application of the YACCA technique to the program from Fig. 3-1 a). In this figure, the names of the constants *M1* and *M2* contain the numbers of the BBs these constants depend on.

```
code = B0;
ERR_CODE = 0;
i = 0;
ERR_CODE |= (code != B0);
code = code ^ (B0 ^ B1);
while(i < n) {
 ERR_CODE |= ((code != B1) && (code != B8)) || (i>=n);
 code = (code & M1_1_8) ^ M2_1_8_2;
 if (a[i] < b[i])
 { ERR_CODE |= (code != B2) || (a[i] >= b[i]);
   code = code ^ (B2 ^ B3);
   x[i] = a[i];
   ERR_CODE |= (code != B3);
   code = code ^ (B3 ^ B4);
 }
 else
 { ERR_CODE |= (code != B2) || (a[i] < b[i]);
   code = code ^ (B2 ^ B5);
   x[i] = b[i];
   ERR_CODE |= (code != B5);
   code = code ^ (B5 ^ B6);
 }
 ERR_CODE |= (code != B4) && (code != B6);
 code = (code & M1_4_6) ^ M2_4_6_7;
 i ++;
 ERR_CODE |= (code != B7);
 code = code ^ (B7 ^ B8);
 }
 ERR_CODE |= ((code != B1) && (code != B8)) || (i<n);
 if(ERR_CODE) error();
```

Figure 3-23. Program hardened according to the YACCA technique

13.2   Experimental results

In order to assess the effectiveness of the proposed approach, several fault injection campaigns were performed and their results were reported in [46]. Experiments were performed using an in-house developed emulation-based fault injection environment [63] on a system composed of a Sparc V8 microprocessor running 4 benchmark programs implementing the following tasks:

- a 5x5 matrix multiplication (M),

- the Kalman Filter (K),
- the fifth order elliptical wave filter (E),
- the Lempel Ziv Welch (LZW) Data Compression algorithm (L).

4 versions for each benchmark were considered:

- an un-hardened version
- a hardened one, obtained applying the CFCSS [44] technique to the original code
- a hardened one, obtained applying the ECCA [45] technique to the original code
- a hardened one, obtained applying the YACCA technique to the original code.

Table 3-5 reports overheads obtained by comparing the size and the execution time of the hardened programs with the original ones. These results demonstrate that the memory and performance overhead caused by the application of the YACCA technique is comparable with the one of the CFCSS technique, but it is always better than that of ECCA technique.

Moreover, the results reported in Table 3-5 show that a large difference in terms of overheads can be obtained considering the different programs. This is due to the different characteristics of the programs' BBs, namely:

- *E* includes several BBs with many mathematical instructions which are CPU intensive, consequently the added instructions are less relevant in terms of size and speed
- *L*, in contrary, includes many BBs with a limited number of instructions.

The results gathered during the fault injection experiments are reported in Tables 3-6 and 3-7, where the transients faults injected in the un-hardened programs are categorized according to their effects and then compared with those injected in the 3 safe versions (CFCSS, ECCA and YACCA).

During experiments randomly selected bit-flips were injected in the immediate operands of the branch instructions, *i.e.*, CFEs of types 1-3 from the fault model presented in section 1 were generated.

Considering the whole set of 16 case studies, the time needed to execute the complete Fault Injection campaign has been about 20 hours.

Table 3-5. Memory and performance overhead numbers for the YACCA technique

Program	Memory overhead [%]			Performance overhead [%]		
	CFCSS	ECCA	YACCA	CFCSS	ECCA	YACCA
M	261	408	191	135	199	147
E	124	153	129	107	120	110
K	164	282	217	117	168	156
L	338	630	496	185	426	354

Fault effects are classified according to Chapter 1. The following acronyms are used:

- *Effect-less:* EL.
- *Fault detected by means of software techniques:* SD.
- *Fault detected by means of EDM:* EDM.
- *Failure:* FA.
- *Time-out:* TO.

The results reported in Tables 3-6 and 3-7 demonstrate the effectiveness of the YACCA method as far as the fault coverage is considered. A very limited number of CFEs cause a failure, and the method shows itself to be more powerful than the considered alternative approaches.

Note that the experimental results obtained considering the CFCSS method present a higher percentage of wrong answers than the one published in [44]. This is mainly due to the following motivations:

1. in experiments performed in [46] the CFCSS technique is applied on the high-level source code, differently from the results published in [44], which are obtained applying the rules on the assembly-level code.
2. Two different fault models are adopted: they present different characteristics, motivating different figures.

Table 3-6. Fault injection experiments (figures are in percentage unless the number of injected faults) for the original programs and programs hardened with YACCA

Prog	Faults [#]	Not hardened				YACCA				
		EL	EDM	FA	TO	EL	SD	EDM	FA	TO
M	5,000	5.5	49.9	20.6	24.0	4.1	56.0	14.2	0.9	24.5
E	5,000	7.8	56.3	10.8	25.1	1.8	54.5	7.6	0.0	35.9
K	5,000	12.3	55.9	11.5	20.4	32.6	22.2	31.5	0.4	13.2
L	1,000	22.3	51.8	25.9	0.0	42.0	21.1	32.1	0.1	4.7

Table 3-7. Fault injection experiments (figures are in percentage unless the number of injected faults) for programs hardened with CFCSS and ECCA

Prog	Faults [#]	CFCSS					ECCA				
		EL	SD	EDM	FA	TO	EL	SD	EDM	FA	TO
M	5,000	3.8	53.5	12.8	19.1	10.5	28.4	49.9	7.3	3.7	10.6
E	5,000	8.8	22.0	33.4	18.7	16.9	24.4	39.8	14.0	4.1	17.5
K	5,000	10.2	42.4	35.2	1.5	10.6	27.3	42.8	21.3	2.2	6.1
L	1,000	17.0	44.5	28.5	6.0	4.0	37.5	42.6	17.6	0.6	1.7

13.3   Advantages and limitations

The method covers all single errors of the types 1-4 from the fault model reported in the section 1. The set and test assertions do not involve divisions or multiplications, so their execution is not time consuming.

To avoid the addition of new branch instructions into the program, which themselves can be the sources of CFEs, the *ERR_CODE* variable may be checked once in the program exit. However, this leads to the introduction of an increased error detection latency. In order to avoid this drawback it is possible to introduce a simple hardware dedicated to performing the test assertions, as described in chapter 5.

14.   SIED AND ITS ENHANCEMENTS

14.1   The approach

The software-based error detection technique proposed in [64][65] and named *Software Implemented Error Detection* (SIED) combines data errors and CFEs detection. Data error detection is based on the instruction duplication, while CFEs are detected by signature-monitoring technique. In [66] a detailed analysis was performed, which allowed identifying and removing the reasons of faults escaping the detection in instruction duplication technique. In [66] a CFC technique is also presented. In this section we present the part of the techniques proposed in [64], [65], [66] referring to CFEs detection; namely, we present the intra-block CFEs detection mechanism proposed in the SIED technique, we discuss the intra-block illegal jumps not detectable by instruction duplication identified in [66], and the solutions proposed in the same work for their detection. Finally, we present the signature-monitoring technique for inter-block CFEs detection presented in [66]. Data errors detection performed through data duplication is presented in details in chapter 2.

14.1.1   Intra-block detection

Intra-block CFE detection is intended to be combined with the instruction duplication technique presented in section 2. For the purpose of intra-block CFE detection a *checkpass* variable is introduced, which is initialized with the value *ni* representing the number of BB's instructions. The *checkpass* variable is decremented after the execution of each original instruction in the program and before the execution of its replica. The variable value is checked on the block's exit. If the *checkpass* variable is not equal to 0 in the end of the program, an error is issued.

As the analysis presented in [66] showed, some intra-block illegal branches still escape detection. The solutions for the detection of these illegal branches suggested in [66] are presented in the following of this subsection.

A *data computing block* (DCB) is a set of sequential instructions that are duplicated for error detection. Each DCB has its workspace, *i.e.*, some resources (registers or variables). Different DCBs may share some resources, for example registers.

A general conclusion made by the authors was that the instruction duplication technique is able to detect faults, which have a single consequence on DCBs (for example, affect either shared or unshared resources of one DCB); faults affecting both shared and unshared resources may escape detection. The main idea of the solutions proposed in [66] for the detection of such faults is to guarantee that at least one of DCB workspaces (the original or its replica) remains undamaged.

Two categories of intra-block CFEs escaping detection were distinguished in [66] basing on their effects:

1. *No dependency with other DCBs*: an intra-block CFE may affect a DCB' (replica of the DCB), when the execution of the DCB is completed corrupting simultaneously the results of the DCB and the DCB'; so that they become incorrect but equal. This situation can happen when the CFE does not permit to entirely complete the DCB' workspace load: as a result, the DCB' workspace contains some values belonging to the DCB, which permit to an error to corrupt both the DCB and DCB' result.

   An example is reported in Fig. 3-24 a). Here *a*1 is an original variable and *a*2 is its replica. The DCB' workspace is not properly loaded due to an illegal jump; consequently, the DCB' uses some content belonging to the DCB. The registers Reg1 and Reg2 are not updated with the correct values: Reg2 is assigned with the address of the *a*1 variable; Reg1 contains the old value of the *a*2 variable, as the new value was not computed due to the error. The old value of *a*2 is incorrectly copied into the *a*1 variable, so the two copies *a*1 and *a*2 contain the same incorrect value, and the consistency check cannot detect an error.

   The solution to this problem proposed in [66] consists in inserting a set of neutralization instructions between the DCB and the DCB'. Neutralization instructions clear the content of all registers used by the DCB before executing the DCB'. In the example from Fig. 3-24 a) two instructions should be added, clearing the registers Reg1 and Reg2, as shown in Fig. 3-24 b).

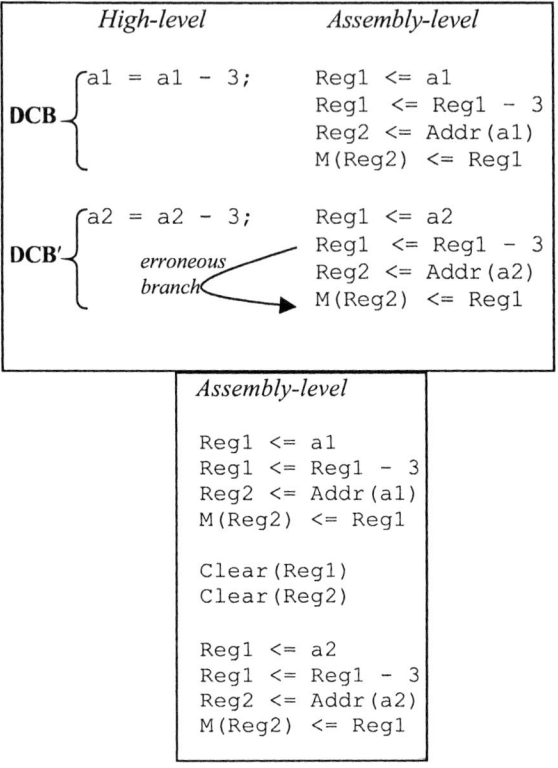

a)   b)

Figure 3-24. Example of undetected intra-DCB fault a),
and possible solution for its detection b)

2. *Dependency with other DCBs*: when a DCB1 is executing an illegal
   branch, it may interrupt the workspace load and transfer the control to
   another block DCB2, which may continue the execution using the
   content of the shared workspace loaded in DCB1.

   For the solution of this problem the authors propose to separate in
   the address space the mutually sensitive DCBs to ensure that there is no
   interaction between mutually sensitive DCBs and their workspaces. The
   solution proposed for single bit-flips is to separate mutually sensitive
   DCBs by a distance *d* so that no single bit-flip error can lead to a jump
   from one to another. The condition on *d* is expressed as follows:

   $$d_{a,b} = b - a \neq 2^i,$$

   where *a* and *b* are physical addresses belonging to blocks DCB1 and
   DCB2, respectively, and *i* is any number from 0 to the program address
   space width.

14.1.2 Inter-block detection

In this section we present the signature-monitoring technique developed for inter-block CFEs detection and presented in [66]. In this technique a unique identification number (named *IDB*) is assigned to each program BB and the checking code is added to the program.

In order to detect not only illegal branches but also wrong branches and in order to avoid signature aliasing, in [66] it is suggested to associate signatures not to BBs but to branches of the CFG. The signature $br_{i,j}$ of the branch $b_{i,j}$ is equal to the concatenation of the identification numbers of the branch source BB v_i and the branch destination BB v_j ($br_{i,j} = IDB_i \mid IDB_j$, where "|" denotes a concatenation operation). The run-time signature B is computed in each BB during the program execution and compared with the branch signature saved in a special variable R in the previously traversed BB.

The checking code (shown in Fig. 3-25) is added at the end of each BB. For simplicity we denoted the concatenation operation in the code with the symbol "|". Firstly, the checking code concatenates the value of the B variable with the *IDB* of the current BB and checks if the value of branch signature B is equal to the signature of the branch saved in the variable R. In case of mismatch an error is detected. Then, the value of R is set to the signature of the next branch to be traversed, and finally the B variable is set to the *IDB* of the current BB. The first two lines of the checking code control that the run-time signature is correct, while the last two lines prepare the transfer to the next BB.

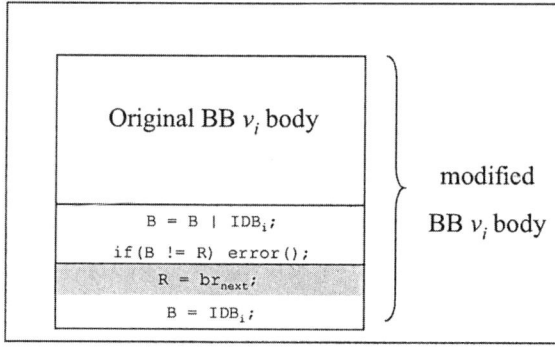

Figure 3-25. BB supplied by checking code

In Fig. 3-25 the instruction shown in gray depends on the type of the branch to be taken from the BB v_i. The authors of the method classified the branches into three types:

- *Certain* – the branch source BB always transfers the control to the same destination BB, and this transition does not depend on any condition.
  Let the branch $b_{i,j}$ be of a certain type. In this case in the branch source BB v_i the value $br_{i,j}$ is assigned to the variable R: $R = br_{i,j}$.

- *Conditional* – the branch source BB transfers the control to two destination BBs: the choice depends on whether a condition is true or false. In this case in order to assign to a variable R the signature of the branch to be taken the additional control of the condition is performed.
  Let BB v_i have two outcoming branches, $i.e.$, $b_{i,j}$ in case the condition is true, and $b_{i,k}$ in case the condition is false. Then, the value is assigned to the variable R in the BB v_i as shown in Fig. 3-26.

```
if(condition)  R = bri,j;
else R = bri,k;
```

Figure 3-26. Variable R assignment in the case of conditional branch

- *Current state dependent* – the branch source BB transfers the control towards two or more BBs depending on the system state. An example of this type of branch is a branch corresponding to the return from the program function, which can be called from different program BBs. In this case a special execution order variable (EO) is introduced in order to predict the correct destination BB.
  Let the predecessors of BB v_i be BBs v_n and v_m; if the BB v_i is reached from BB v_n, then from the BB v_i the CF transfers to the BB v_k, while in case it is reached from the BB v_m the CF transfers to the BB v_l. Then, the BBs v_n and v_m assign to variable EO the values 1 and 2, correspondingly, and in the BB v_i the value is assigned to the variable R as it is presented in Fig. 3-27.

```
if(EO == 1)  R = bri,k;
if(EO == 2)  R = bri,l;
```

Figure 3-27. Variable R assignment in the case of current state dependant branch

In order to detect illegal branches inside a BB or branches corrupting the checking instructions a *local cumulative signature N* is introduced. For each BB $N = N1 + N2 + N3 = 0$, where $N1$, $N2$ and $N3$ are unique for each BB. The position of the components of N in the BB is shown in Fig. 3-28.

Finally, in order to avoid any erroneous program interruption or program reset, which could lead to CFE escaping, the authors introduce the START BB, which can be executed only once and the STOP BB with a special signature, which is reproduced only if the program completed correctly.

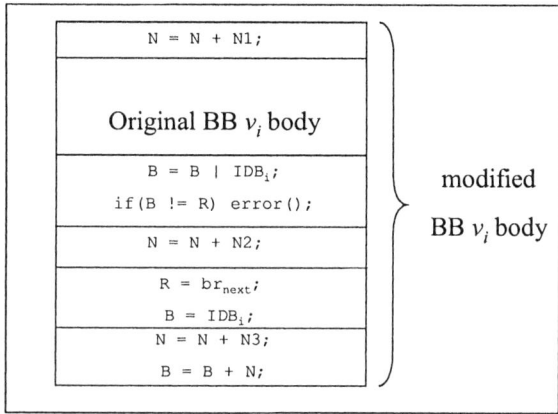

Figure 3-28. BB complemented by the checking code including the local cumulative signature checking

14.2   Experimental results

In order to evaluate the method some experiments were performed, and their results reported in [66], where 3 synthetic (*i.e.*, specially developed) programs and 3 real applications were hardened with the proposed technique. Both data errors and CFEs detection techniques were implemented. For the considered applications the observed execution time increase was about three times, while the program size increase was about four times.

Exhaustive fault injection campaigns were performed on two processors: LEON and a digital signal processor. During the fault injection experiments bit-flips were injected at all possible processor cycles in all bits of the general-purpose registers, the program stack, the pointer register, and the program counter register. For each application some hundred thousand bit-flips were injected. All experiments showed zero undetected faults.

14.3   Advantages and limitations

The method described in [66] is able to cover all single inter-block CFEs of types 1-4 from the fault model reported in section 1. The method considers intra-block CFEs detection. Moreover, in [66] border cases, which are able to cause CFE escaping are analyzed and solutions are proposed to overcome the CFE escaping. Experimental results performed by authors of the method in [66] report a 100% detection of the injected bit-flip faults.

The main limitation of the method lies in the significant memory overhead: although the checking operations introduced by the method do not involve such time consuming operations as division or multiplication, their number is significant (see Fig. 3-28). Besides, some parts of the approach are suitable to be applied to high-level descriptions of the program, while others to assembly-level, which complicates the approach implementation. Some details of the method are not explicitly described, for example the implementation of the START and STOP BBs.

15.   REFERENCES

42. G. Miremadi, J. Karlsson, U. Gunneflo, J. Torin, "Two Software Techniques for On-line Error Detection", Digest of Papers of the Twenty-Second International Symposium on Fault-Tolerant Computing, 8-10 July 1992, pp. 328 – 335.
43. S.S. Yau, F.-C. Chen, "An Approach to Concurrent Control Flow Checking", IEEE Transactions on Software Engineering, Vol. 6, No. 2, March 1980, pp. 126-137.
44. N. Oh, P.P. Shirvani, E.J. McCluskey, "Control-Flow Checking by Software Signatures", IEEE Transactions on Reliability, Vol. 51, No. 2, March 2002, pp. 111-122.
45. Z. Alkhalifa, V.S.S. Nair, N. Krishnamurthy, J.A. Abraham, "Design and Evaluation of System-Level Checks for On-Line Control Flow Error Detection", IEEE Transactions on Parallel and Distributed Systems, Vol. 10, No. 6, June 1999, pp. 627-641.
46. O. Goloubeva, M. Rebaudengo, M. Sonza Reorda, M. Violante. "Soft-Error Detection Using Control Flow Assertions", Proceedings of the 18th International Symposium on Defect and Fault Tolerance in VLSI Systems, 3-5 November 2003, pp. 581-588.
47. O. Goloubeva, M. Rebaudengo, M. Sonza Reorda, M. Violante. "Improved Software-Based Processor Control-Flow Errors Detection Technique", Proceedings of the Annual Reliability and Maintainability Symposium, 26-29 January 2005, pp. 583-589.
48. C.H. Tung, C.W. McCarron, "Concurrent Control Flow Checking in Sequential and Parallel Programs", Conference Record of the Twenty-Fourth Asilomar Conference on Signals, Systems and Computers, Vol. 2, 5-7 November 1990, pp. 851 – 855.

49. G. Miremadi, J. Torin, "Evaluating Processor-Behavior and Three Error-Detection Mechanisms Using Physical Fault-Injection", IEEE Transactions on Reliability, Vol. 44, No. 3, September 1995, pp. 441 – 454.

50. R.G. Halse, C. Preece, "Erroneous Execution and Recovery in Microprocessor Systems", Software and Microsystems, Vol. 4, No. 3, June 1985, pp. 63-70.

51. G.A.S. Wingate, C. Preece, "Performance Evaluation of a new Design Tool for Microprocessor Transient Fault Recovery", Microprocessing and Microprogramming, Vol. 27, 1989, pp. 801-808.

52. M.A. Schuette, J.P. Shen, "Exploiting instruction-level parallelism for integrated control-flow monitoring", IEEE Transactions on Computers, Vol. 43, No. 2, February 1994, pp. 129 - 140.

53. P. Furtado, H. Madeira, "Fault Injection Evaluation of Assigned Signatures in RISC Processors", Proceedings of the 2nd European Dependable Computing Conference, 1996, pp. 55-72.

54. J. Carreira, H. Madeira, J.G. Silva, "Xception: Software Fault Injection and Monitoring in Processor Functional Units", Proceedings of the 5th Conference on Dependable Computing for Critical Applications, 27-29 September 1995.

55. G.A. Kanawati, V.S.S. Nair, N. Krishnamurthy, J.A. Abraham, "Evaluation of Integrated System-Level Checks for On-Line Error Detection", Proceedings of the IEEE International Computer Performance and Dependability Symposium, 4-6 September 1996, pp. 292 – 301.

56. Z. Alkhalifa, V.S.S. Nair, "Design of a Portable Control-Flow Checking Technique", Proceedings of the High-Assurance Systems Engineering Workshop, 11-12 August 1997, pp. 120 – 123.

57. G.A. Kanawati, N.A. Kanawati, J.A. Abraham, "FERRARI: a Flexible Software-Based Fault and Error Injection System", IEEE Transactions on Computers, Vol. 44, No. 2, February 1995, pp. 248-260.

58. M. Rebaudengo, M. Sonza Reorda, M. Torchiano, M. Violante, "Soft-Error Detection through Software Fault-Tolerance Techniques", Proceedings of the IEEE International Symposium on Defect and Fault Tolerance in VLSI Systems, 1-3 November 1999, pp. 210-218.

59. P. Cheynet, B. Nicolescu, R. Velazco, M. Rebaudengo, M. Sonza Reorda, M. Violante, "Experimentally Evaluating an Automatic Approach for Generating Safety-Critical Software with Respect to Transient Errors", IEEE Transactions on Nuclear Science, Vol. 47, No. 6, December 2000, pp. 2231-2236.

60. A. Aho, R. Sethi, J. Ullman, "Compilers: Principles, Techniques and Tools", Addison-Wesley, 1986.

61. A. Benso, S. Di Carlo, G. Di Natale, P. Prinetto, L. Tagliaferri, "Control-Flow Checking Via Regular Expressions", Proceedings of the IEEE Asian Test Symposium, 19-21 November 2001, pp. 299-303.

62. R. Venkatasubramanian, J.P. Hayes, B.T. Murray, "Low-cost On-line Fault Detection Using Control Flow Assertions", Proceedings of the International On-Line Testing Symposium, 7-9 July 2003, pp.137 – 143.

63. P.L. Civera, L. Macchiarulo, M. Rebaudengo, M. Sonza Reorda, M. Violante, "Exploiting Circuit Emulation for Fast Hardness Evaluation", IEEE Transactions on Nuclear Science, Vol. 48, No. 6, December 2001, pp. 2210 –2216.

64. B. Nicolescu, Y. Savaria, R. Velazco, "SIED: Software Implemented Error Detection", Proceedings of the 18th IEEE International Symposium on Defect and Fault Tolerance in VLSI Systems, 3-5 November 2003, pp. 589 – 596.

65. B. Nicolescu, Y. Savaria, R. Velazco, "Performance Evaluation and Failure Rate Prediction for the Soft Implemented Error Detection Technique", Proceedings of the 10th IEEE International On-Line Testing Symposium, 12-14 July 2004, pp. 233 – 238.

66. B. Nicolescu, Y. Savaria, R. Velazco, "Software Detection Mechanisms Providing Full Coverage Against Single Bit-Flip Faults", IEEE Transactions on Nuclear Science, Vol. 1, No. 6, December 2004, pp. 3510 – 3518.

Chapter 4

ACHIEVING FAULT TOLERANCE

1.   INTRODUCTION

In this chapter the main techniques to harden an unreliable system and transform it into a fault-tolerant one are presented.

When fault tolerance (and not only fault detection) capabilities are the target, the approaches presented so far are not enough: some of them can be extended (at a higher cost in terms of memory, performance, and development cost) to cope with the more stringent requirements. New approaches can be devised, coping with these requirements. Obviously, the same assumptions holding for the previous chapters are valid here: therefore, we will mainly focus on techniques allowing to reach the target (i.e., fault tolerance) resorting only to changes in the software (while the hardware is not affected); moreover, we will focus mainly on techniques whose adoption can be automated easily. The techniques that are covered in this chapter are design diversity, checkpointing, algorithm-based fault tolerance, and duplication.

2.   DESIGN DIVERSITY

The concept of design diversity is very old. At the beginning of the XIX century Charles Babbage, known as the "Father of Computing", has suggested that "the most certain and effectual check upon errors which arise in the process of computation is to cause the same computation to be made by separate and independent computers; this check is rendered still more decisive if they make their computations by different methods" [67]. This theory can be transferred and adapted easily to the modern computer science.

The use of redundant copies of hardware, data and programs' instruction has proven to be quite effective in the detection of physical faults and in subsequent system recovery. However, design faults – which are introduced by human mistakes or defective design tools – are reproduced when redundant copies are made. *Design diversity* is the approach in which the hardware and software elements that are to be used for multiple computations are not copied, but are independently designed to fulfill the same function through implementations based on different technologies. A definition of design diversity has been given in [68] as "production of two or more systems aimed at delivering the same service through separate designs and realizations".

Design diversity is the common technique adopted to achieve software fault tolerance. Two or more versions of software are developed by independent teams of programmers and software engineers, and by using different techniques, methodologies, algorithms, programming languages and programming compilers. However, all the different implementations of the software meet the common requirements and specifications.

The versions of software produced through the design diversity approach are called *variants* (or *versions* or *alternates*). Besides the existence of at least two variants of a system, tolerance of faults needs a *decider* (or *acceptance test*), aimed at providing an error-free result from the variants execution; the variants execution have to be performed from consistent initial conditions and inputs. The common specification has to address explicitly the *decision points* defined as:
• the time when the decisions have to be performed
• the data processed by the decider.

The two most common techniques implementing design diversity are *N-Version Programming (NVP)* [69] and *Recovery Blocks (RB)* [70]. These techniques have mainly been introduced to face the effects of software bugs. However, they can also be adopted to address hardware faults; they do not depend on any particular error model and are able to detect (and in some cases correct) both transient and permanent errors. A deeper analysis will be given in the following.

Design diversity intrinsically exploits code replication and introduces a high overhead in terms of memory area and performance slow-down. Hardened versions, based on design diversity, and focused on fault detection, only, require doubling the memory area and the elapsed time. On the other hand the fault-tolerant version requires more than 3 times than the memory area occupied by the original version and lasts more than 3 times than the time required by the un-hardened version.

2.1   N-version programming

N-version programming requires the separate, independent preparation of multiple (i.e., N) versions of a program for some application. These versions are executed in parallel. At the system-level an application environment controls their execution. Each receives identical inputs, and each produces its version of the required outputs. A voter collects the outputs that should, in principle, all be the same (*consensus*). If the outputs disagree, the system detects the error and can tolerate it using the results of the majority, provided there is one.

N-version programming is easily classified as a *static redundant scheme* and presents many analogies with the triple modular redundancy (TMR) and the N-modular redundancy (NMR) approach used for tolerating hardware failures [71].

The decision mechanism is defined as follows:

- A set of program state variables are to be included in a *comparison vector* (c-vector); each program stores its c-vector.
- The N programs possess all the necessary attributes for concurrent execution.
- At a specified *cross-check point* each program generates its comparison vector and a decider executes the decision algorithm comparing the c-vectors and looks for the *consensus* of two or more c-vectors among the N versions.

The N independent versions of the software can be run on a single computer, one after another, or alternatively they could be run simultaneously on independent computers.

Software diversity may be specified in the following elements of the design process:

1. training, experience, and location of implementing personnel
2. application algorithms and data structures
3. programming languages
4. software development methods
5. programming tools and environments
6. testing methods and tools.

The purpose of such required diversity is to minimize the opportunities for common causes of faults in two or more versions.

This method has been widely exploited to target possible software errors, i.e., design faults or software bugs[72], but can be effectively adapted to detect possible hardware errors. A transient error in a hardware component has the same effect as a software one. If any of the hardware components experiences an error, it causes the software version running on that hardware to produce inaccurate results. If all of the other hardware and software modules are functioning properly, the system will still produce a correct result, since the majority (e.g., 2 out 3 versions) is correct. If a hardware component and the software component running on it both have errors, the system can again continue to correctly function, if all the other hardware and software components are functioning properly. If less than $\lceil N/2 \rceil$ software or hardware components behave correctly, the system may fail, since the majority does not produce the correct result. The Fault-Tolerant Processor-Attached Processor (FTP-AP) architecture proposed in [73] may be seen as an implementation of this hardware-software fault-tolerant architecture. A quadruple configuration of a core architecture is used as a support for the management of the execution of 4 diversified software modules running on 4 distinct processors.

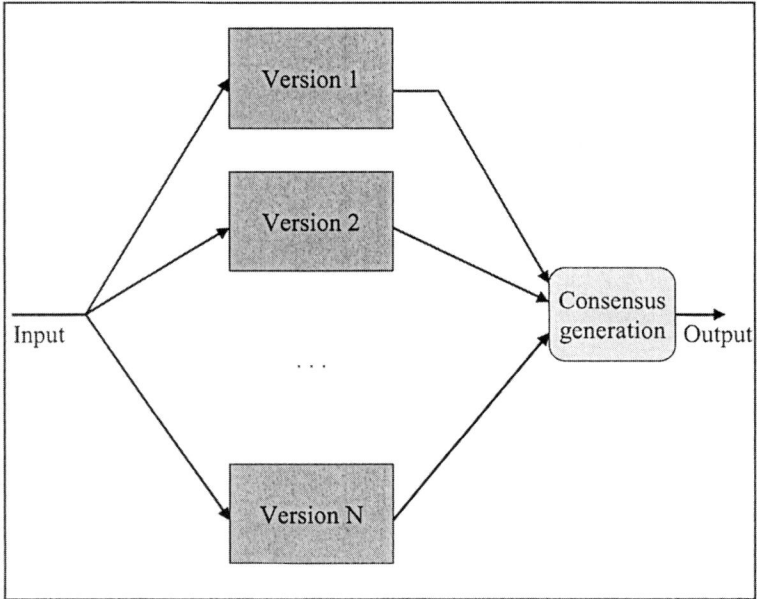

Figure 4-1. N-Version programming.

N-Version programming has been exploited in many industrial applications: NASA Space Shuttle [74], Airbus A320/A330/A340 [75] and Boeing 777 aircraft control [76] and various railway signaling and control systems [77] [78] [79].

2.1.1 Time redundancy

Virtual duplex systems (VDS) described in Section 2.3 can be extended from only detecting faults to tolerating faults, also, using three versions of a software with identical functionalities. Two versions are used to detect transient faults, the third is needed for recovery.

A fault tolerant VDS system using a microprocessor that supports multiple threads in hardware is presented in [37]. The system is composed of 3 versions of a software with identical functionalities. Two versions are used to detect transient faults, the third will be needed for detection of permanent faults and for recovery. The versions are built through design diversity to be able to recover from transient as well as from many permanent hardware faults.

A fault tolerant VDS exploits simultaneous multithreading in hardware: 2 threads execute in parallel a particular version. At regular times (called *rounds*) the versions are compared and a state is saved in the form of a checkpoint (see more details about checkpoint in the next Section 4.3). If the states disagree a fault is detected. After the detection of a fault, while the first thread executes version 3 for i rounds, the other thread is used to proceed versions 1 and 2 beyond round i (*roll-forward scheme*). In particular, in order to detect a fault, versions 1 and 2 started from a common state. Here there is the possibility to choose from the states P or Q of both versions at the end of round i, respectively. However, these states are different, and it is unknown which of these states is affected by the fault just detected.

In a probabilistic scheme, a state is chosen randomly, and both versions are executed for $i/2$ rounds each, which needs the same time as executing i rounds of version 3 in the first thread. A comparison is made through a majority voting among 3 states (states of version 1, 2 and 3 at round i). If the chosen initial state is the state of the fault-free version, the roll-forward is successful, and, after the majority voting, the process re-starts from round $i+i/2$; otherwise the roll-forward does not give any gain, and the process re-starts again from round i. Since the choice is random, the probability to choose correct version is 0.5. If a particular part of hardware is more likely to be affected by faults, it is possible to use some prediction scheme, which might increase this probability (exploiting techniques similar to branch prediction and keeping a history of faults).

In a deterministic scheme, first $i/4$ rounds of version 2 are executed starting from state P (the state of version 1 after round i), then $i/4$ rounds of version 1 are executed starting from state P, then $i/4$ rounds of version 1 are executed starting from state Q (the state of version 2 after round i), and finally $i/4$ rounds of version 2 are executed starting from state Q. With this scheme the roll-forward is always successful, and after the majority voting, the process re-starts always from round $i+i/4$.

To complete recovery in case of a successful roll-forward, the state of the fault-free version (version 1 or 2) is copied to version 3. So, version 3 is rolled forward to the fault-free version and forms a new VDS with the remaining fault-free version.

In order to maximize the efficiency of the roll-forward scheme in the probabilistic scheme 3 multithreads are needed executing versions 1 and 2 for i rounds each in 2 separate threads and version 3 in another thread; in the deterministic scheme 5 multithreads are needed executing versions 1 and 2 for i rounds starting from P and Q each in 4 separate threads and version 3 in a another thread.

2.2   Recovery Block

Another major evolution of hardware and software fault-tolerance has been the recovery block (RB) approach [70].

Recovery block exploits software redundancy. The recovery block scheme consists of 3 software elements:
1. a *primary* module which normally executes the critical software function
2. an *acceptance test* which checks the outputs for correctness
3. an *alternate*[2] module which performs the same function as the primary module, and is invoked by the acceptance test upon detection of a failure in the primary module.

In this approach these elements are organized in a manner similar to the passive dynamic redundancy (*standby sparing*) technique adopted for the hardware fault tolerance. The recovery block approach attempts to prevent software faults from impacting on the system environment, and it is aimed at providing fault-tolerant functional components which may be nested within a sequential program. The usual syntax is shown in Fig. *2-6*.

---

[2] The term *alternate* reflects sequential execution, which is a feature specific to the recovery block approach.

```
ensure Acceptance Test
by primary alternate
else by alternate 2
else by alternate 3
else by alternate 4
...
else by alternate N
else error
```

Figure 4-2. The syntax of the Recovery Block scheme.

On entry to a recovery block the state of the system must be saved to permit rollback error recovery. RB performs run-time software, as well as hardware, error detection by applying the acceptance test to the outcome delivered by the primary alternate. If the acceptance test is passed, the outcome is regarded as successful and the recovery block can be exited, discarding the information on the state of the system taken on entry. However, if the acceptance test is not passed (or if any errors are detected by other means during the execution of an alternate), recovery is implemented by state restoration: the system rolls back and starts executing an alternate module from the previously established correct intermediate point or system state, known as *recovery point*. Recovery is considered complete when the acceptance test is passed or all the modules are exhausted. If all the alternates either fail the test or result in an exception (due to an internal error being detected), a failure exception is signaled to the environment of the recovery block. Since the recovery block can be nested, then the raising of such an exception from an inner recovery block would invoke recovery in the enclosing block.

Fig. *4-3* shows a scheme of the Recovery Block approach.

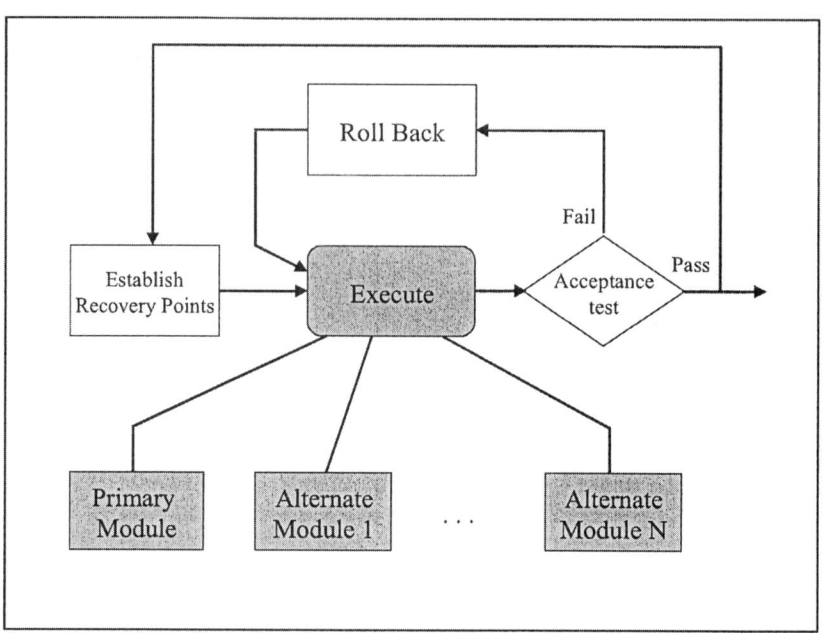

Figure 4-3. Recovery Block approach.

In general, multiple alternate procedures can be used. Each procedure must be deliberately designed to be as independent as possible, so as to minimize the probability of having correlated errors in the primary and in the alternate modules. This may be achieved by enforcing design diversity with independently written program specifications, different program languages, algorithms, etc, as described in Section 4.2.

The acceptance test must be simple, otherwise there will be a significant chance that it itself contains a fault, and so fails to detect some errors, and/or identifies falsely some conditions as being erroneous. Moreover, the test introduces a run-time overhead which could be unacceptable. A number of possible methods for designing acceptance test have been proposed (more details can be found in [81]) but none has been defined as the golden method. Generally speaking, the application test is dependent on the application. As an example, in [83] Algorithm-Based Fault Tolerance (ABFT) error detection techniques are exploited to provide cheap and effective acceptance tests. ABFT (more details will be provided later in this chapter) has been used in numerical processing for the detection of errors. ABFT technique provides a transparent error checking method embedded into the functional procedure that can be effectively applied in a recovery block scheme. This method can be applied whenever ABFT is applicable.

Indeed in many real-time applications, the majority of which involve control systems, the numerical processing involved can be adapted to an ABFT solution.

Although each of the alternates within a recovery block has to satisfy the same acceptance test, there is no requirement that they all must produce the same results. The only constraint is that the results must be acceptable as determined by the test. Thus, while the primary alternate should attempt to produce the desired outcome, the further alternate may only attempt to provide a degraded service. This is particularly useful in real-time systems, since there may be insufficient time available for complete functional alternates to be executed when a fault is encountered. The extreme case corresponds to a recovery block which contains a primary module and a null alternate. Under these conditions, the role of the recovery block is simply to detect and recover from errors.

In the normal, and most probable case, only the primary alternate of the recovery block is executed as well as the acceptance test, and the run-time overhead of the recovery block is kept to a minimum.

2.2.1 Distributed Recovery Block

The Distributed Recovery Block (DRB) scheme [84] is an approach for achieving both hardware and software fault tolerance in real-time distributed and/or parallel computer systems.

The underlying design philosophy behind the DRB scheme is that a real-time distributed [85] or parallel [86] computer system can take the desirable modular form of an interconnection of *computing stations*, where a computing station refers to a processing node dedicated to the execution of one or a few application tasks.

In a basic configuration, a *computing station* consists of two *self-checking processing nodes* (PSP) executing functionally equivalent tasks, the first node being called the *primary node* and the second node being called the *shadow node*. Each PSP possesses the capability of judging the reasonableness of its task execution results through a software acceptation test or a hardware self-checking circuit.

In the following description we will consider the general case that the arrival rate of data is such that data may arrive when other data are still being processing. In order to manage this general case it is thus necessary to provide input data queues in each node within a PSP station. Each node may contain multiple input data queues corresponding to multiple data sources. Therefore, it is important for the partner nodes in a PSP station to ensure that they process the same data item in each task execution cycle. This is achieved by associating an identifier (ID) to each data.

The schema is organized as follows:
- Both nodes (primary and shadow) obtain input data from a multicast channel
- The primary node informs the shadow node of the ID of the data item that the former received for processing in the current task cycle
- The primary and shadow nodes process the data item and perform their self-checking concurrently by using the same acceptance test routine
- Since the primary node passes the test, it delivers the results to both the successor computing station(s) and the shadow node, and then starts the next task cycle
- By receiving the output from the primary node, the shadow node detects the success of the primary node and, if the shadow node also succeeded in its acceptance test, it too starts the next task cycle.

Fig. *4-4* shows a fault-free task execution cycle in a PSP station.

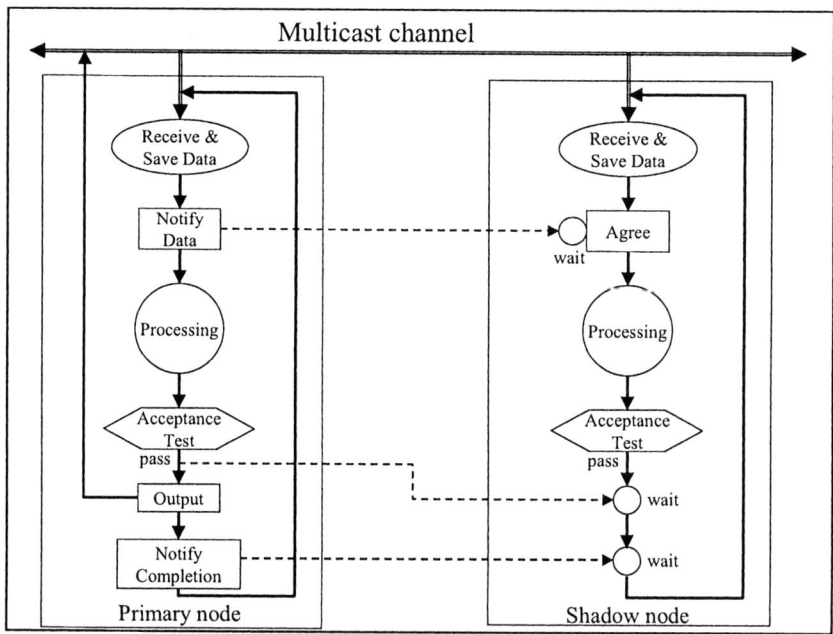

Figure 4-4. A fault-free task execution cycle of a PSP station.

Let suppose the following faulty case:
- The primary node fails in passing the acceptance test or crashes during the processing of the data item whereas the shadow node passes
- The shadow node then learns the failure of the primary node by noticing the absence of output from the primary node

- The shadow node then becomes a new primary and delivers its task execution results to both its successor computing station(s) and the primary node
- Meanwhile, the primary node, if alive, attempts to become a new useful shadow node by making a retry of the processing of the saved data item. If the primary node passes the acceptance test this time, it can then continue as a useful shadow node and proceeds to the next task cycle.

Fig. *2-8* shows a task execution cycle of a PSP station involving a failure.

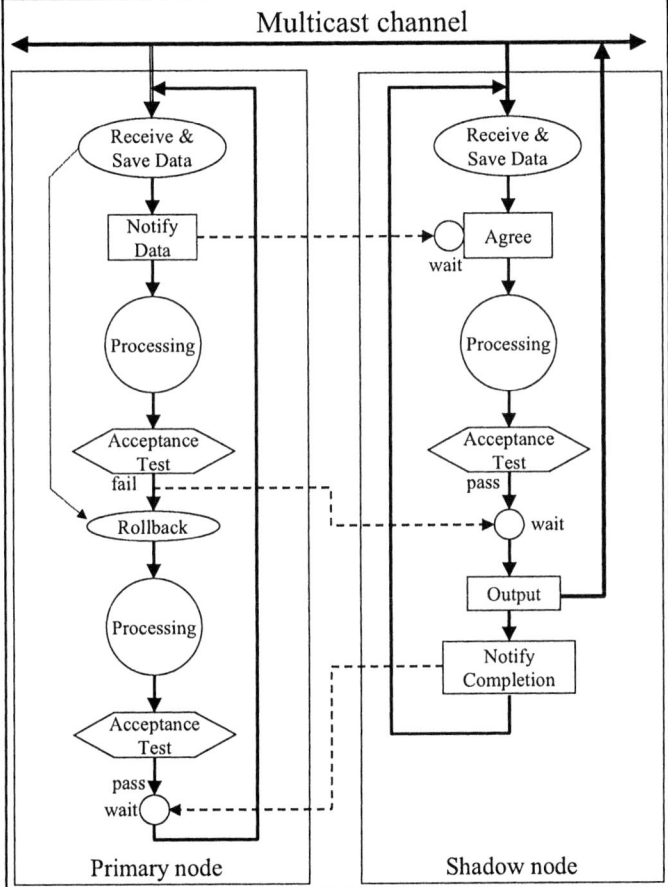

Figure 4-5. A task execution cycle of a PSP station involving a failure.

In order to support not only hardware faults, but also software faults, the above primary-shadow PSP scheme can be extended by incorporating the approach of using multiple versions of the application task procedure. Such versions are called *try blocks*. The extended scheme is the *Distributed Recovery Block* (DRB) scheme and it uses the recovery block language construct to support the incorporation of try blocks and the acceptance test. Let consider the Recovery Block schema reported in Fig. *2-6*, the syntax of recovery block is shown in Fig. *4-6*, where T denotes the acceptance test, B_1 the primary try block, and B_k (with $2 \leq k \leq n$), the *alternate try blocks*.

All the try blocks are designed the produce the same or similar computational results. The acceptance test is a logical expression representing the criterion for determining the acceptability of the execution results of the try blocks. The execution of a try block is thus always followed by an acceptance test. If an error is detected during the execution of a try block or as a result of an acceptance test execution, then a rollback-and-retry with another try block follows. A try not completed within the maximum execution time allowed for each try block due to hardware faults or excessive looping is treated as a failure.

In the DRB scheme, a recovery block is replicated into multiple nodes forming a DRB computing station for parallel redundant processing.

In most cases a recovery block contains just 2 try blocks. With this configuration, the roles of two try blocks are assigned differently in the two nodes. The governing rule is that the primary node tries to execute the primary try block whenever possible whereas the shadow node tries to execute the alternate try block.

The fault-free execution observes the following steps:
- Both nodes receive the same input data,
- They process the data by use of two different try blocks
- They check the results by use of the acceptance test concurrently.

If the primary node fails and the shadow node passes its own acceptance test, the shadow immediately delivers its processing results to the successor computing stations. The two nodes then exchange their roles, i.e., the shadow assumes the primary's role.

If the shadow node fails, the primary node is not disturbed. Whichever node fails, the failed node attempts to become an operational shadow node without disturbing the (new) primary node; it attempts to roll back and retry with its second try block to bring its application computation state updated.

```
ensure T
by B₁
else by B₂
...
else by Bₙ
else error
```

Figure 4-6. Distributed Recovery Block Scheme.

A distributed fault tolerant system for process control based on Distributed Recovery Block has been implemented and integrated into a chemical processing control system [87].

3.   CHECKPOINTING

Checkpointing is a commonly used technique for reducing the execution time for long-running programs in the presence of failures. With checkpointing the status of the program under execution is saved intermittently in a reliable storage. Upon the occurrence of a failure, the program execution is restarted from the most recent checkpoint rather than from the beginning.

In checkpointing schemes the task is divided into n intervals. At the end of each interval a checkpoint is added, either by the programmer [88] or by the compiler [89-90]. Fault detection is obtained exploiting hardware redundancy by duplicating the task into two or more processors and comparing the states of the processors at the checkpoints. The probability of two faults resulting in identical states is negligible, so that two matching states indicate a correct execution. By saving at each checkpoint the state of the task in a reliable storage, the need to restart the task after each fault is avoided[3]. Instead, the task can be rolled back to the last correct checkpoint, and execution resumed from there, thereby shortening fault recovery. Reducing the task execution time is very important in many applications like real-time systems, with hard deadlines, and transactions systems, where high availability is required.

---

[3] Task duplication [91] was introduced to detect transient faults, based on duplicating the computation of a task on two processors. If the results of the two executions do not match, the task is executed again in another processor until a pair of processors produces identical results. This scheme does not use checkpoints, and every time a fault is detected the task has to be started from its beginning.

Different recovery techniques are used to shorten the fault recovery time:

1. *rollback recovery* [88]: both processors are set back to the state of the last checkpoint and the processing interval is retried. If two equal states are reached afterwards, the processing is continued

2. *stop and retry recovery* [88]: if a state comparison mismatches, both processors are stopped until a third processors computes a third status for the mismatching round. Then a 2-out-of-3 decision is made to identify the fault free version that is used to continue duplex processing

3. *roll-forward checkpoint* [91]: if a state comparison mismatches, the two different states are both stored. The state at the preceding checkpoint, where both processing modules had agreed, is loaded into a spare module and the checkpoint interval is retried on the spare module. Concurrently, the task continues forward on the two active modules, beyond the checkpoint where the disagreement occurred. At the next checkpoint, the state of the spare module is compared with the stored states of the two active modules. The active module, which disagrees with the spare module, is identified to be faulty and its state is restored to the correct one by copying the state from the other active module, which is fault free. The spare is released after recovery is completed. The spare can be shared among many processor pairs and used temporarily when fault occurs.

In checkpointing schemes a checkpointing overhead is introduced due to the time to store the processors' states and the time to compare these states. The time spent for compare and store operations may vary significantly, depending on the system, and thus the checkpointing overhead is determined mainly by the operation that takes a longer time. As an example, in a cluster of workstations connected by a LAN, the bandwidth of the communications subsystem is usually lower than the bandwidth of the local storage subsystem. On the other hand, in multiprocessor supercomputers without local disks at the computing nodes, the bandwidth of the communication subsystem is usually higher than the bandwidth of the local storage subsystem.

Different methods have been proposed to reduce checkpointing overhead. The first method is to tune the scheme to the specific system that is implemented on, and use both the compare and the state operations efficiently [92]. Using two types of checkpoint (compare-checkpoints and store-checkpoints) allows tuning the scheme to the system. The compare-checkpoints are used to compare the states of the processors without storing them, while in the store-checkpoints the processors store their states without comparison. Using two types of checkpoints enables choosing different frequencies for the two checkpoint operations, and utilizing both operations in an efficient way. When the checkpoints that are associated with the operation that takes less time are used more frequently than the checkpoints associated with the operation that takes more time, the recovery time after fault can be reduced without increasing the checkpoint overhead. This leads to a significant reduction in the average execution time of a task.

The second method is to reduce the comparison time by using signatures [91], instead of comparing the whole states of the processors. In systems with high comparison time, signatures can significantly reduce the checkpoint overhead, and hence reduce the execution time of a task.

The tradeoffs involved in choosing an appropriate checkpoint frequency are the following. Very frequent checkpoints cause high overhead due to checkpointing durations, while too rare checkpoints cause longer fault latency and may cause a more probable failure. The effects of varied check intervals and checkpoint periods have been studied in [92]. A main result from that study is that shortening test intervals improves dependability, because the likeliness of two processes affected by a fault is decreased. Thus, it is advised to test states more often than saving checkpoints.

4. ALGORITHM-BASED FAULT TOLERANCE (ABFT)

This technique has been first proposed in [93], and then improved and extended in several papers appeared in the following years.

4.1 Basic technique

In its basic version, the technique presented by Huang and Abraham [93] in 1984 is aimed at hardening processors when executing matrix applications, such as multiplication, inversion, LU decomposition. Hardening is obtained by adding coding information to matrices: however, while other approaches introduce coding information (to detect and possibly correct errors) to each byte or word, these coding information are added to whole data structures (in this case to each matrix) or, according to the authors definition, at the *algorithm level*.

4.2 Matrix multiplication

4.2.1 Method description

The algorithm is based on modifying the matrices the application is working on according to the definitions introduced in the following.

Definition 1

Given a matrix A composed of n × m elements, the corresponding *column checksum matrix* A_c is an (n+1) × m matrix, which consists of the matrix A in the first n rows and a column summation vector in the (n+1)-th row. Each element of the column summation vector corresponds to the sum of the elements of the corresponding column (see Fig. *4-7*.a). An example of a 3 × 3 column checksum matrix is reported in Fig. *4-8*.

Definition 2

Given a matrix A composed of n × m elements, the corresponding *row checksum matrix* A_r is an n × (m+1) matrix, which consists of the matrix A in the first m columns and a row summation vector in the (m+1)-th column. Each element of the row summation vector corresponds to the sum of the elements of the corresponding row (see Fig. *4-7*.b). An example of a 3 × 3 row checksum matrix is reported in Fig. *4-9*.

Definition 3

Given a matrix A composed of n × m elements, the corresponding *full checksum matrix* A_f is an (n+1) × (m+1) matrix, which is the column checksum matrix of the row checksum matrix A_r of A (see Fig. *4-7*.c). An example of a 3 × 3 full checksum matrix is reported in Fig. *4-10*.

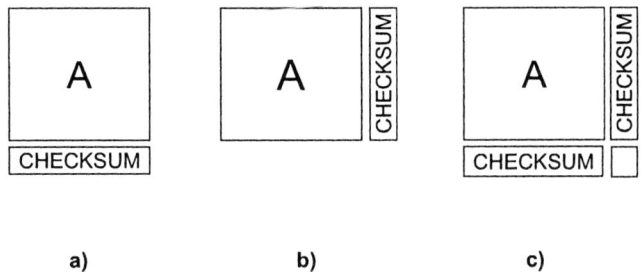

Figure 4-7. A column (a), row (b) and full (c) checksum matrix.

Figure 4-8. A 3 × 3 integer matrix and the corresponding column checksum matrix.

Figure 4-9. A 3 × 3 integer matrix and the corresponding row checksum matrix

1	2	3
4	5	6
7	8	9

1	2	3	6
4	5	6	15
7	8	9	24
12	15	18	45

Figure 4-10. A 3 × 3 integer matrix and the corresponding full checksum matrix

The technique proposed in [93] is based on the observation that some matrix operations (matrix by matrix multiplication, LU decomposition, addition, matrix by scalar multiplication, transposition) preserve the checksum property, according to the following theorems (whose proof can be found in [93]).

Theorem 1

When multiplying a column checksum matrix A_c by a row checksum matrix B_r, the result is a full checksum matrix C_f. Moreover, the following relation holds among the corresponding information matrices:

$A * B = C$

Fig. *4-11* shows how the ABFT technique implements matrix multiplication, while Fig. *4-12* gives an example.

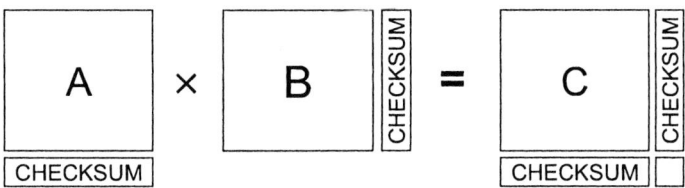

Figure 4-11. Multiplication according to the ABFT technique

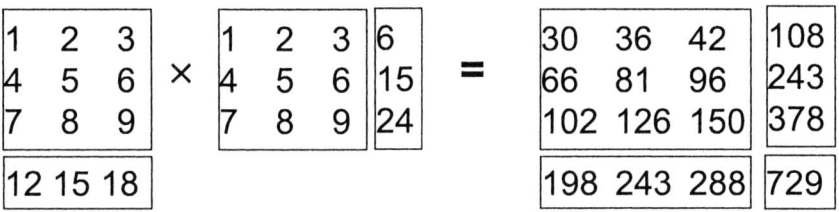

Figure 4-12. Example of matrix multiplication according to the ABFT technique

Theorem 2
When a matrix C is LU decomposable, the corresponding full checksum matrix C_f can be decomposed into a column checksum lower matrix and a row checksum upper matrix.

Theorem 3
When adding two full checksum matrices A_f and B_f, the result is a full checksum matrix C_f. Moreover, the following relation holds among the corresponding information matrices:
A + B = C

Theorem 4
The product of a full checksum matrix and a scalar value is a full checksum matrix.

Theorem 5
The transpose of a full checksum matrix is a full checksum matrix.

In order to harden an application performing a matrix operation, one can therefore proceed as follows:

- The operand matrices are transformed into the corresponding row, column, or full checksum matrices, depending on the operation
- The operation is performed on the checksum matrices
- A check is performed to detect possible errors, corresponding to the following steps:
  - The sum of all the elements on each row and column is computed
  - The resulting value is compared with that stored in the row or column summation vector; if a difference is observed, an error is detected

If we assume that the detected error affected a single element in the result matrix, the identification of the affected element can be performed resorting to the following sequence of operations:

- If a mismatch on both a row and a column summation vector element is detected, the error affected the information element at the intersection of the inconsistent row and column (Fig. *4-13*)
- If a mismatch is detected on a row or column summation vector element, only, the error affected the summation vector (Fig. *4-14*).

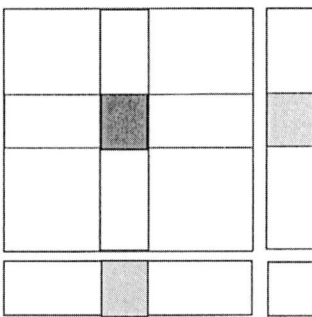

Figure 4-13. Faulty matrix element identification

Figure 4-14. Faulty column summation element identification

After the identification of the faulty element, its correction can be performed resorting to the following sequence of operations:
- If the error affected an information element, the error can be corrected by computing its fault-free value subtracting the sum of the values of the other elements on the same row or column from the corresponding element in the row or column summation vector

• If the error affected an element of a row or column summation vector, the fault free value of element can be computed by adding all the elements of the row or column.

4.2.2   Comments

It is important to note that in the case of matrices composed of floating point elements, roundoff errors could create problems to comparison operations. In this case, some false alarms could be raised. A method to compute the thresholds to be used for distinguishing between roundoff errors and errors stemming from faults is outlined in [94] for a similar case.

The ABFT technique is particularly attracting because it introduces a memory and performance overhead that, when compared with other techniques (e.g., TMR), is relatively limited. Since the introduced memory overhead meanly corresponds to an additional row and column and it grows linearly with the matrix size as $O(N)$, but the percentage overhead decreases when the matrix size increases, because the memory size grows as $O(N^2)$.

The error detection and correction capabilities of the method are very high when faults affecting the matrices elements during the computation are considered. The method is able to detect and correct any error affecting a single element in the final matrix. On the other side, the correction capabilities are limited if an error affects more than one element in the resulting matrix.

On the other side, the method is rather weak in detecting and correcting other kinds of faults, e.g., those affecting the memory elements in a microprocessor control unit. If errors in the application code are considered, the method shows some detection capabilities, corresponding to data alterations, although it is definitely unable to detect all the errors of this category.

4.3   FFT

The Algorithm-Based approach has been extended to other problems: one of them is the Fast Fourier Transform.

The method has been introduced in [95], where an ad hoc hardware architecture was supposed to be adopted: the original target was to modify the algorithm to detect faults arising in this architecture. However, the method is suitable to be adopted even if the FFT algorithm is implemented in software on a conventional computer.

4.3.1   Method description

The discrete Fast Fourier Transform of a sequence x(n) can be computed as

$$X(k) = \sum_{n=0}^{N-1} x(n) w_N^{kn}, \quad k = 0,1,\ldots, N-1$$

where

$$w_N = e^{-j(2\pi/N)}$$

The computation can be performed either on a standard computer, resorting to a matrix recording all the required products $x(n)w^{kn}$, or on a special purpose architecture (named *FFT network*), whose architecture is shown in Fig. *4-15*.

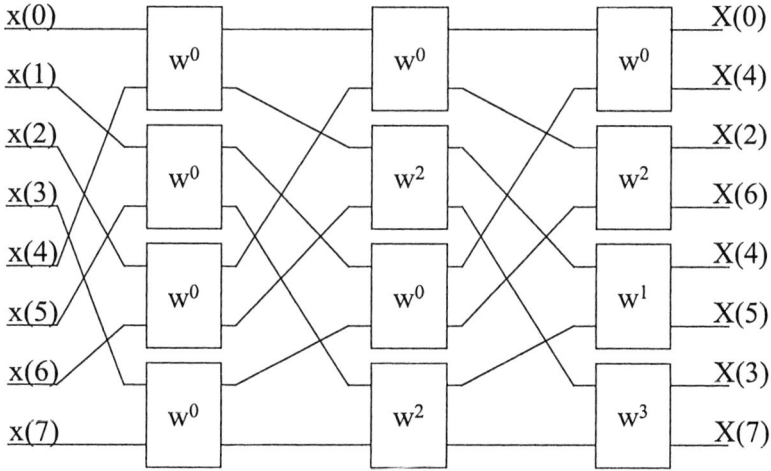

Figure 4-15. FFT network architecture

Each rectangle corresponds to a two-point butterfly (Fig. *4-16*), whose two outputs implement the following functions:

$$c = a + b * w_N^k$$
$$d = a - b * w_N^k$$

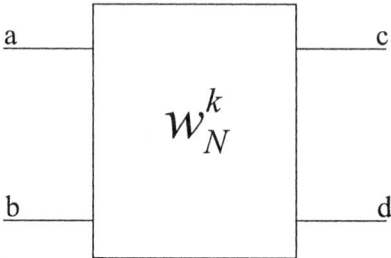

Figure 4-16. Two-point butterfly

In order to harden this architecture against possible faults affecting the composing modules, Jou and Abraham proposed to encode the inputs, so that a checksum can be computed out of the outputs. The proposed technique is based on substituting each input x with $ax + bx^1$, where a and b are properly chosen integer constants, and x^1 is the element of the input sequence, rotated by one position. With this encoding, the k-th output y_k must be decoded by multiplying it by a factor equal to

$$\frac{1}{a + bw_N^{-k}}$$

The correct outputs must fulfill the following relation

$$Nx(0) = \sum_{n=0}^{N} \frac{y_k}{a + bw_N^{-k}}$$

This approach leads to the modified FFT network shown in Fig. *4-17* (where a=2 and b=1).

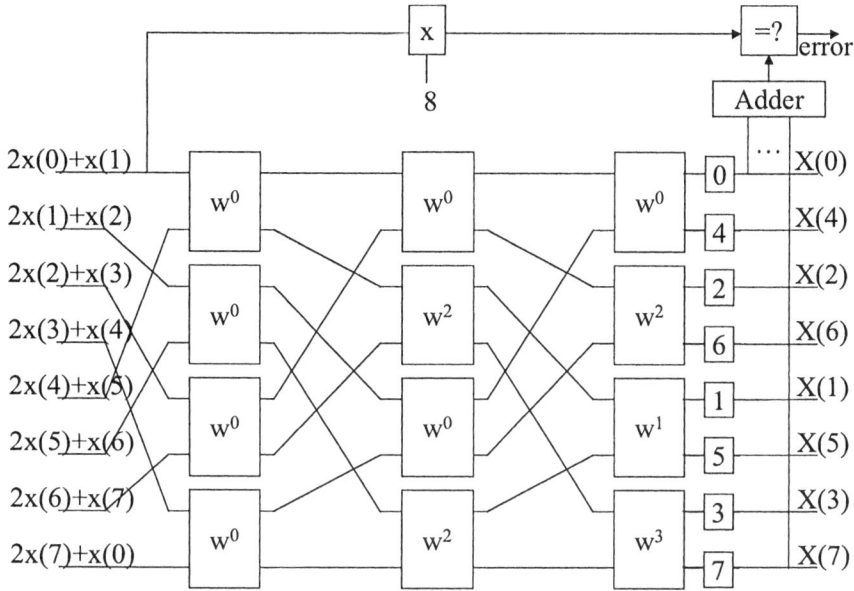

Figure 4-17. FFT network architecture

In their paper Jou and Abraham demonstrate that using this encoding, a very high percentage (greater than 99%) of faults affecting both the data and the computation elements can be detected. They also propose a method to identify the faulty component, that can be used to reconfigure the network, and hence lead to a fault tolerant system.

The same approach can be adopted if the discrete FFT is computed in software on a conventional architecture.

4.4 Final comments

The algorithm-based approach attracted a lot of interest in the last two decades, and has been widely adopted to several common problems with good results in terms of detection capabilities, and relatively low requirements in terms of memory and performance overhead.

An important limitation of the algorithm-based approach is that it can only be applied to those algorithms for which an ABFT version has been devised, mainly correspondent to regular data structures. Moreover, it requires properly modifying the application algorithm in order to implement the fault tolerant version, thus making impossible to reuse existing libraries.

As discussed for the FFT algorithm, the approach can be extended to the case of non-conventional architectures executing the application (this case is not covered in this book) [95][96][97].

5.   DUPLICATION

The technique is based on a set of transformation rules applied to a high-level code in order to obtain a

5.1   Duplication and checksum

The method first proposed in [98] and then fully described in [99] extends the one proposed by the same authors in [100], in such a way that not only detection, but also fault tolerance is achieved.

The method focuses on computing-intensive applications, only. Therefore, it is assumed that the program to be hardened begins with an *initialization phase*, during which the data to be elaborated are acquired. This phase is then followed by a *data manipulation phase*, where an algorithm is executed over the acquired data. At the end of the computation, the computed results are committed to the program user, through a *result presentation phase*. The proposed code transformation rules are meant to be applied on the algorithm executed during the data manipulation phase.

The approach exploits code transformation rules providing fault detection and, for most cases, fault correction. The rules are intended for being automatically applied to the program source high-level code and can be classified in two broad categories: rules for detecting and correcting faults affecting data and rules for detecting and (when possible) correcting faults affecting code.

5.1.1   Detecting and correcting transient faults affecting data

Data hardening is performed according to the following rules:
- Every variable x must be duplicated: let x_0 and x_1 be the names of the two copies. Every write operation performed on x must be performed on x_0 and x_1. Two sets of variables are thus obtained, the former (set 0) holding all the variables with footer 0 and the latter (set 1) holding all the variables with footer 1.
- After each read operation on x, the two copies x_0 and x_1 must be checked, and if an inconsistency is detected a recovery procedure is activated.

- One checksum c associated to one set of variables is defined. The initial value of the checksum is equal to the exor of all the already initialized variables in the associated set.
- Before every write operation on x, the checksum is re-computed, thus canceling the previous value of x ($c' = c \wedge x_0$).
- After every write operation on x, the checksum is updated with the new value x' ($c' = c \wedge x'_0$).

The recovery procedure re-computes the exor on the set of variables associated to the checksum (set 0, for example), and compares it with the stored one. Then, if the re-computed checksum matches the stored one, the associated set of variables is copied over the other one; otherwise the second set is copied over the first one (e.g., set 0 is copied over set 1, otherwise set 1 is copied over set 0).

In order to provide a sample example of how the proposed method works, let us consider the code fragment reported in Fig. *4-18*. When all the proposed rules are applied, the hardened code is the one reported in Fig. *4-19*. In Fig. *4-19*, function chk() computes the exclusive-or of all the variables in the set 0.

```
int a, b;
    ...
a = b;
```

Figure 4-18. Original code fragment

```
int a0, a1, b0, b1, c;
  ...
c = c^a0;
a0 = b0;
a1 = b1;
c = c^a0;    /* c is updated */
if(b0!=b1) /* error detection */
  if(chk()==c) /* error correction */
  { b1 = b0; /* b1 is wrong */
    a1 = a0; /* a1 is wrong */
  } else
  { b0 = b1; /* b0 is wrong */
    a0 = a1; /* a0 is wrong */
    c = chk();
  }
```

Figure 4-19. Hardened code fragment

5.1.2   Detecting and correcting transient faults affecting the code

To detect faults affecting the code the method exploits the techniques introduced in [101]. The first technique consists in executing any operation twice, and then verifying the coherency of the resulting execution flow. Since most operations are already duplicated due to the application of the rules described in the previous sub-section, this idea mainly requires the duplication of the jump instructions. In the case of conditional statements, this can be accomplished by repeating twice the evaluation of the condition.

The second technique aims at detecting those faults modifying the code so that incorrect jumps are executed, resulting in a faulty execution flow. This is obtained by associating an identifier to each *basic block* in the code. An additional instruction is added at the beginning of each block of instructions. The added instruction writes the identifier associated to the block in an ad hoc variable, whose value is then checked for consistency at the end of the block.

The recovery procedure consists in a rollback scheme: as soon as a fault affecting the program execution flow is detected, the program is restarted (i.e., the program execution is restarted from the data manipulation phase, or from a safe point which has been previously recorded). Thanks to this solution, we are able to:

*   *Detect and correct* transient faults located in the processor internal memory elements (e.g., program counter, stack pointer, stack memory elements) that temporarily modify the program execution flow.

- *Detect* transient faults originated in the processor code segment (where the program binary code is stored) that permanently modify the program execution flow. As soon as a SEU hits the program code memory, the bit-flip it produces is indeed permanently stored in the memory, causing permanent modification to the program binary code. Restarting the program execution when such a kind of fault is detected is insufficient for removing the fault from the system. As a result, the program enters in an end-less loop, since it is restarted every time the fault is detected. This situation can be easily identified by a watch-dog timer that monitors the program operations.

5.1.3 Results

In [99] the authors report some experimental results allowing to evaluate the advantages and disadvantages of their method.

Starting from a set of benchmark programs, they first obtained their fault tolerant versions by applying the proposed source code transformation rules. For this purpose they exploited an extended version of the tool presented in [102]. Then they evaluated the area overhead introduced by the method by measuring the size of the code and data segments of the fault tolerant versions and by relating them with those of the unhardened ones. They also measured the time overhead the method introduces, as the ratio between the number of clock cycles needed for executing the fault tolerant programs and the unhardened ones. Finally, they evaluated the error detection and correction capabilities of the method by performing simulation-based fault injection experiments on an Intel 8051-based system. During the experiments, they injected randomly selected (both in time and space) bit-flips in the program *data segment*, storing the data the program manipulates and the stack, and in the *code segment* storing the binary code the processor executes. For each benchmark, the authors executed a preliminary set of fault injection experiments to measure the impact of faults in the unhardened program; then they executed a new set of fault injection experiments on the fault tolerant version of the same program: 10,000 random faults were injected in each experiment. Fault injection experiments were performed resorting to the emulation-based environment presented in [103].

Faults have been classified according to the categories described in Section 1.2.3.

In the experiments three programs were considered: *Sieve* implements the sieve of Eratosthenes over a set of N_0 bytes; *Bubble sort* implements the Bubble sort algorithm over a set of N_1 integers; *Matrix* implements the product of two $N_2 \times N_2$ matrices of integer numbers. The adopted processor core implements the Intel 8051 instruction set and includes a 128-bytes internal memory. Moreover, it is able to run programs up to 1,024-bytes long. Given these constraints the following set of parameters were adopted for the considered benchmark programs: $N_0=40$, $N_1=10$, $N_2=2$. While evaluating the overhead introduced by the approach with respect to the unhardened version, the authors recorded the figures reported in Table *4-1*.

Table 4-1. Data, code and performance overheads.

	Data segment size increase	Code segment size increase	CPU time increase
Sieve	2.2	2.1	2.7
Bubble sort	2.2	2.8	1.8
Matrix	2.2	3.8	2.4

To compare the figures of Table *4-1* with a reference approach, in [99] the authors report a comparison with the figure obtained with a software TMR version of the considered benchmarks. Data reported for this version show an average data segment overhead of 3.5, an average code segment increase of 3.0 and an average performance overhead of 3.1. As a result, the authors of [99] state that the proposed approach is able to provide fault tolerance while reducing the memory overhead with respect to the TMR approach, while the performance penalties introduced by the two methods are comparable.

The results gathered during fault injection experiments are reported in Tables *4-2* and *4-3*, where transients faults injected in the unhardened programs are categorized according to their effects and then compared with those injected in the fault tolerant versions. The figures show that the proposed method is able to significantly improve the error detection and correction capabilities of a given applications. As far as faults inside the data segment are considered, the method provides complete fault coverage: the number of *failures* is indeed always reduced to 0 for the hardened versions. The same result was observed when faults affecting the code segment were analyzed, where *failures* are reduced to 0 in all the considered programs. From Tables *4-2* and *4-3*, one can also observe that many faults exist that can only be *detected*. Most of them are provoked by SEUs hitting the memory area storing the result of the program at the very beginning of the program execution. Furthermore, many faults hitting the code area are classified as *time-out*. These are faults that let the program enter in an endless loop and that trigger the watch-dog timer embedded in the fault injection system.

Table 4-2. Fault injection in data segment of the Intel 8051-based system

	Sieve		Bubble sort		Matrix	
	Original	Hardened	Original	Hardened	Original	Hardened
Injected	10,000	10,000	10,000	10,000	10,000	10,000
Effect-less	8,294	6,487	9,227	8,058	9,398	8,213
Corrected	0	813	0	1,568	0	653
Failure	1,701	0	773	0	580	0
Software Detected	0	2,697	0	374	0	283
Time-out Detected	5	3	0	0	22	855

Table 4-3. Fault injection in code segment of the Intel 8051-based system

	Sieve		Bubble sort		Matrix	
	Original	Hardened	Original	Hardened	Original	Hardened
Injected	10,000	10,000	10,000	10,000	10,000	10,000
Effect-less	9,041	5,465	9,136	6,048	8,944	6,763
Corrected	0	498	0	585	0	314
Failure	416	0	637	0	579	0
Software Detected	0	192	0	10	0	130
Time-out Detected	543	3,845	227	3,357	487	2,792

5.2   Duplication and Hamming code

The method exploit's the properties of the error correcting codes for achieving fault-tolerance. Error correcting code introduces information redundancy into the source code. These transformations, which were automatically performed, introduce data and code devoted to detect, and eventually correct, possible errors corrupting information stored in the memory area.

The method has been proposed by Nicolescu et al. in [107]. The basic idea is to associate an extra code information to every variable in the code. This extra code (*Hamming corrector*) is computed according to Hamming codes. This code is able to correct a single error and to detect a double error. A detailed description of how this code is computed is out of the scope of this book.

Every time a variable is modified, its correspondent Hamming corrector has to be updated. On the other hand, in a read operation the Hamming corrector is used for the decoding operation. Two parameters are needed for this operation: the variable's value and the variable's hamming corrector code. In the case of the corruption of one of these two values, the decoding procedure will take one of the possible following decisions:

• If one bit is corrupted, the decision is a correction of the corrupted bit

- It two bits are damaged, then the decision is the detection, without correction possibility
- If more than two bits are affected, the decision is an erroneous correction.

Original code	Modified Code
| ```
...
a = 5;
...
b = a + 2;
...
``` | ```
a = 5;
a_code = code (a);
...
b = decode (a_code,a) + 2;
b_code = code (b);
...
``` |

Figure 4-20. Hamming code-based redundancy.

Fig. *4-20* hows an example for simple piece of code of the resulting program including Hamming codes.

Experimental results executed injecting faults into the memory and internal registers of a RISC processor (transputer T225) demonstrate the feasibility of the approach and its detection and correction capability. As far as fault affecting data are considered 0,7% of faults produce a failure, but 36% of faults are detected and 32% of faults are detected and corrected. As far as faults affecting code are considered, 3% of faults produce a failure, but 53% of them are detected and 0,2% are detected and corrected.

The major drawback of error detection and correction methods based in Hamming codes comes from the resulting memory area overhead (due to hamming corrector codes and decoding operations) and the increase in execution time due to hamming corrector code update and decoding operation execution. The overhead factors obtained considering a benchmark program corresponds to a execution time 12 times than the one required for the unhardened program and a memory size 3 times than the one for the original unhardened program. These overhead factors show that this method can be applied where the fault tolerance requirements justify those high overhead penalty.

REFERENCES

67. C. Babbage, "On the mathematical powers of the calculating engine," unpublished manuscript, December 1837, Oxford, Buxton Ms7, Museum of History of Science.

Printed in "The Origins of Digital Computers: Selected Papers", B. Randell (ed.), Springer, 1974, pp. 17-52

68. A. Avizienis, J.C. Laprie, "Dependable Computing: from concepts to design diversity," Proceedings of the IEEE, Vol. 74, No. 5, May 1986, pp. 629-638

69. A. Avizienis, "The N-Version approach to fault-tolerant software," IEEE Transactions on Software Engineering, Vol. 11, No. 12, December 1985, pp. 1491-1501

70. B. Randell, "System Structure for Software Fault Tolerance," IEEE Trans. on Software Engineering, Vol. 1, No. 2, June 1975, pp.220-232

71. D. Pradhan, "Fault-tolerant Computer System Design", Prentice Hall, 1996

72. J. P. Kelly, T. I. McVittie, W.I. Yamamoto, "Implementing design diversity to achieve fault tolerance", IEEE Software, Vol. 8, no. 4, July 1991, pp. 61-71

73. J. H. Lala, L.S. Alger, "Hardware and Software Fault Tolerance: a unified Architectural Approach", Proceedings of the 18-[th] International Symposium on Fault-Tolerant Computing, FTCS-18, 1988, pp. 240-245

74. C. E. Price, "Fault tolerant avionics for the Space Shuttle" Proceedings of the 10-[th] IEEE/AIAA Digital Avionics Systems Conference, 1991, pp. 203-206

75. D. Briere, P. Traverse, "AIRBUS A320/A330/A340 Electrical Flight Controls: A Family of Fault-Tolerant Systems", Proceedings of the 23-[rd] International Symposium on Fault-Tolerant Computing, FTCS-23, 1993, pp. 616-623

76. R. Riter, "Modeling and testing a critical fault-tolerant multi-process system", Proceedings of the 25-[th] International Symposium on Fault-Tolerant Computing, FTCS-25, 1995, pp. 516-521

77. G. Hagelin, "ERICSSON safety system for railway control", Proceedings of the Workshop on Design Diversity in Action, Springer Verlag, 1988, pp 11-21

78. H. Kanzt, C. Koza, "The ELEKTRA railway signalling system: field experience with an actively replicated system with diversity", Proceedings of the 25-th International Symposium on Fault-Tolerant Computing, FTCS-25, 1995, pp. 453-458

79. A. Amendola, L. Impagliazzo, P. Marmo, G. Mongardi, G. Sartore, "Architecture and safety requirements of the ACC railway interlocking system", Proceedings of IEEE International Computer Performance and Dependability Symposium, 1996, pp. 21 – 29

80. B. Fechner, J. Keller, P. Sobe, "Performance estimation of virtual duplex systems on simultaneous multithreaded processors", 18-th International Parallel and Distributed Processing Symposium, 2004, pp. 214-217

81. K. Echtle, B. Hinz, T. Nikolov, "On Hardware Fault Detection by Diverse Software", Proceedings of the 13-th International Conference on Fault-Tolerant Systems and Diagnostics," 1990, pp. 362-367

82. T. Anderson, P.A. Lee, Fault Tolerance: Principles and Practice, Prentice Hall, 1981

83. A.M. Tyrrell, Recovery blocks and algorithm-based fault tolerance, EUROMICRO 96. 'Beyond 2000: Hardware and Software Design Strategies', Proceedings of the 22nd EuroMicro Conference,1996, pp. 292 – 299

84. K.H. Kim, H.O. Welch, "Distributed Execution of Recovery Blocks: an approach to uniform treatment of Hardware and Software Faults in Real-Time Applications", IEEE Transactions on Computers, May 1989, pp. 626-636

85. K.H. Kim, L. Bacellar, C. Subbaraman, "Primary-shadow consistency issues in the DRB scheme and the recovery time bound", Proceedings of the 7-th International Symposium on Software Reliability Engineering, 1996, pp. 319-329

86. K.H. Kim, A. Kavianpour, "A distributed recovery block approach to fault-tolerant execution of application tasks in hypercubes", IEEE Transactions on Parallel and Distributed Systems, Vol. 4 , No. 1, Jan. 1993, pp. 104-111

87. M. Hecht, J. Agron, H. Hecht, K.H. Kim, "A distributed fault tolerant architecture for nuclear reactor and other critical process control applications", Proceedings of the 21-st International Symposium on Fault-Tolerant Computing, 1991, FTCS-21, pp. 462-498

88. K.M. Chandy, C.V. Ramamoorthy, "Rollback and recovery strategies for computer programs," IEEE Transactions on Computers, Vol. 21, No. 6, June 1972, pp. 546-556

89. W.K, Fuchs, C.-C. J. Li, "CATCH - compiler-assisted techniques for checkpointing," Proceedings of the 20-th International Symposium on Fault-Tolerant Computing, FTCS-20, 1990, pp. 74-81

90. J. Long, W.K, Fuchs, J.A. Abraham, "Compiler-Assisted Static Checkpoint insertion," 22-nd International Symposium on Fault-Tolerant Computing, (FTCS-22), 1992, pp. 58-65

91. D. K. Pradhan, N. H. Vaidya, "Roll-Forward Checkpointing Scheme: A Novel Fault-Tolerant Architecture," IEEE Transactions on Computers, Vol. 43, No. 10, October 1994, pp. 1163-1174

92. A. Ziv, J. Bruck, "Performance Optimization of Checkpointing Scheme with Task Duplication," IEEE Transactions on Computers, Vol. 46, No. 12, December 1997, pp. 1381- 1386

93. K. H. Huang, J. A. Abraham, "Algorithm-Based Fault Tolerance for Matrix Operations", IEEE Transactions on Computers, vol. C-33, No. 6, June 1984, pp. 518-528

94. A. Roy-Chowdhury, P. Banerjee, "Tolerance Determination for Algorithm Based Checks using Simplified Error Analysis", Proc. IEEE International Fault Tolerant Computing Symposium, 1993

95. J.-Y. Jou, J.A. Abraham, "Fault-Tolerant FFT Networks", IEEE Transactions on Computers, Vol. 37, No. 5, May 1988, pp. 548-561

96. S.-J- Wang, N.K. Jha, "Algorithm-Based Fault Tolerance for FFT Networks", IEEE Transactions on Computers, Vol. 43, No. 7, July 1994, pp. 849-854

97. A. Mishra, P. Banerjee, "An Algorithm-Based Error Detection Scheme for the Multigrid Method", IEEE Transactions on Computers, Vol. 52, No. 9, September 2003, pp. 1089-1099

98. M. Rebaudengo, M. Sonza Reorda, M. Violante, "A New Software-based technique for low-cost Fault-Tolerant application", IEEE Annual Reliability and Maintainability Symposium, 2003, pp. 25-28

99. M. Rebaudengo, M. Sonza Reorda, M. Violante, "A new approach to software-implemented fault tolerance", JETTA: The Journal of Electronic Testing: Theory and Applications, Kluwer Academic Publishers, N. 20, August 2004, pp. 433-437.

100. P. Cheynet, B. Nicolescu, R. Velazco, M. Rebaudengo, M. Sonza Reorda, M. Violante, "Experimentally evaluating an automatic approach for generating safety-critical software with respect to transient errors", IEEE Transactions on Nuclear Science, Vol. 47, No. 6, December 2000, pp. 2231-2236

101. O. Goloubeva, M. Rebaudengo, M. Sonza Reorda, M. Violante, "Soft-error Detection Using Control Flow Assertions", DFT2003: IEEE International Symposium on Defect and Fault Tolerance in VLSI Systems, 2003, pp. 581-588

102. M. Rebaudengo, M. Sonza Reorda, M. Torchiano, M. Violante, "A source-to-source compiler for generating dependable software", Proceedings of the IEEE International Workshop on Source Code Analysis and Manipulation (SCAM), 2001, pp. 33-42

103. P. Civera, L. Macchiarulo, M. Rebaudengo, M. Sonza Reorda, M. Violante, "An FPGA-based approach for speeding-up Fault Injection campaigns on safety-critical circuits", Journal of Electronic Testing: Theory and Applications (JETTA), Kluwer Academic Publishers, Vol. 18, No. 3, June 2002, pp. 261-271

104. O. Goloubeva, M. Rebaudengo, M. Sonza Reorda, M. Violante, "Software Techniques for Dependable Computer-based Systems", chapter in Space radiation environment and its effects on spacecraft components and systems, Cépaduès éd., Toulouse (France), ISBN 2-85428-654-5, 2004

105. M. Rebaudengo, M. Sonza Reorda, M. Torchiano, M. Violante, "An experimental evaluation of the effectiveness of automatic rule-based transformations for safety-critical applications", DFT'00, IEEE International Symposium on Defect and Fault Tolerance in VLSI Systems, October 2000, pp. 257-265

106. M. Turmon, R. Granat, D.S. Katz, J.Z. Lou, "Tests and Tolerances for High-Performance Software-Implemented Fault Detection", IEEE Transactions on Computers, Vol. 52, No. 5, May 2003, pp. 579-591

107. B. Nicolescu, R. Velazco, M. Sonza Reorda, "Effectiveness and Limitations of Various Software Techniques for "Soft Error" Detection: a comparative study", Proceedings of the IEEE 7-th International On-Line Testing Workshop, 2001, pp. 172-177

Chapter 5

HYBRID TECHNIQUES

1. INTRODUCTION

Although this book is devoted to methods aimed at reaching safety and fault tolerance through software techniques, we decided to allocate at least one chapter to hybrid methods, i.e., to those methods that combine changes in the application code with some sort of external (with respect to the processor executing the code) hardware support. The reason for this choice is that for most of the methods presented in this chapter the changes required in the hardware are limited to adding some special device (often named *watchdog*), which interacts with the processor, and checks for possible errors, possibly exploiting special instructions that have been added in the code to support this interaction. In this way, a mix of hardware and software techniques is exploited, resulting in systems having either a higher reliability, or a lower overhead than for those exploiting purely software hardening techniques.

The operation of a watchdog is a two phase process. In the first, the *initialization* phase, the watchdog is provided with a *reference* information about the fault-free operation of the checked processor. In the second one, the *checker* phase, the reference information is compared to the *run-time* information collected by the watchdog processor concurrently. In the case of a discrepancy, an error is detected. The scheme is the one of general testing: the watchdog compares the run-time information from the processor (device under test) with the reference one; the result of the comparison is an error signal. Watchdog devices are generally connected to the bus as shown in Fig. *5-1*. They either simply monitor the bus, or interact with the processor via special commands sent by the processor (which sometimes sees them as I/O devices, i.e., as ports associated to some address). Sometimes, watchdog devices are able to execute a program, which is related to that executed concurrently by the main processor. In this case watchdog devices take the name of watchdog processors (or *coprocessor*).

Historically, the watchdog processor is an extension of the idea of *watchdog timers* ([147]), that are simple hardware or software modules used to monitor concurrently timing (duration) of selected system activities. The system is designed such that under normal operation it signals the watchdog timer within a specified time interval. The signal presets the timer to the initial value. The timer generates an error if no preset signal is received during the specified time interval. Many malfunctions can occur while the system still generates a correct timing signal, and so this approach is usually combined with others to increase the percentage of detected errors.

Hardening performed by resorting to watchdog devices is often categorized as system-level hardening, since it works at the application level, thanks to the interaction with the application software run by the main processor. This kind of hardening can obviously be combined with others (e.g., possible hardening techniques applied within the processor).

Many of the current superscalar processors include features, called Performance Monitoring features, to measure and monitor various parameters related to the performance of the processors. The Performance Monitoring features use special internal counters, which can be initialized to count the occurrences of several events in the processor. Examples of such events are cache hits, instructions executed, and branches taken. Some processors have also special pins, called event-ticking pins, which can signal out the occurrence of internal events of processors. Performance Monitoring features are exploited in [159] for developing a watchdog system (more details are given later).

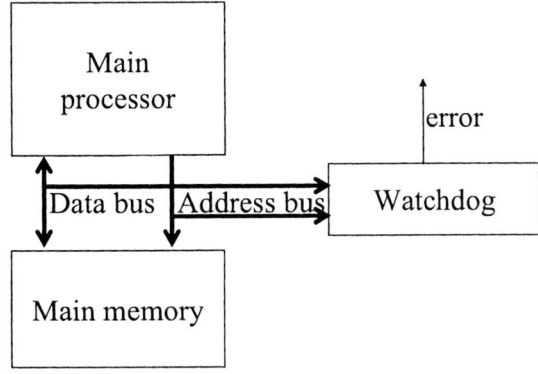

Figure 5-1. Typical architecture of a system including a watchdog

The classification adopted in this chapter is mainly that introduced by [108], which has been further extended to cope with some recently introduced approaches. The adopted classification relates to the kind of checks performed by the watchdog, and thus on the kind of errors that can be detected by it.

2. CONTROL FLOW CHECKING

The methods belonging to this category are strictly connected with those described in Chapter 3, where a watchdog has been added with the purpose of either increasing the number of detected faults, or (more often) to decrease the overhead in terms of performance degradation.

All the methods belonging to this category adopt the concept of *node*, i.e., a group of instructions (corresponding to a single statement, a basic block, a loop-free interval, or others, depending on the method). At the compile time, the source code is divided into nodes, and a *signature instruction* is embedded into the block (at the beginning and/or at the end according to the method). The signature instruction has a field that contains an identifying opcode, and a field that contains the reference signature, as shown in Fig. 5-2. The opcode could be a coprocessor opcode already included in the processor's instruction set, or it could be a specific addition to the instruction set. During the run-time execution, the watchdog observes the executed instructions and generates each node's run-time signature using dedicated hardware. When a signature instruction is detected, the processor may execute a *No-operation* (NOP) instruction, while the watchdog compares the run-time and reference signature, signaling an error if they differ.

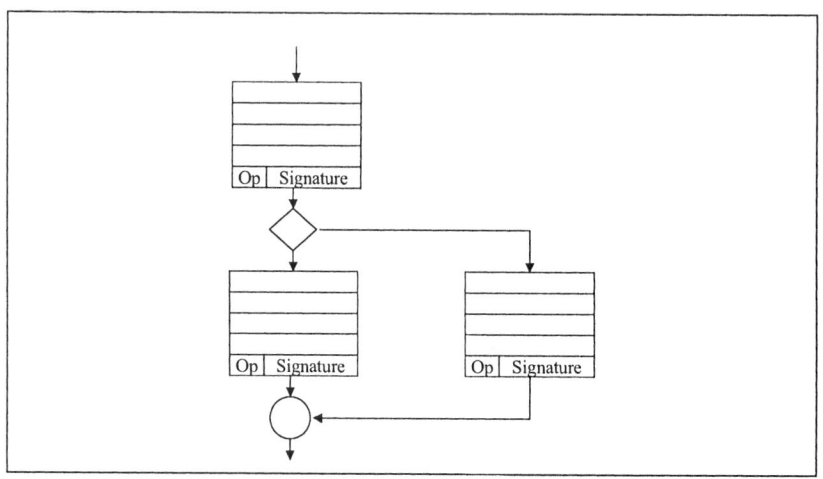

Figure 5-2. Basic signature monitoring technique.

Methods differ in the definition of the node, in the mechanism adopted for computing the associated signature, and in the way the watchdog monitors the control flow.

Two different approaches can be distinguished:
- *Assigned run-time signatures*: the signatures labeling the nodes are assigned arbitrarily (e.g., using prime numbers or successive integers). These signatures are transferred to the watchdog explicitly by the checked processor. Signature transfer statements are inserted at the compile time into the source of the checked program.
- *Derived run-time signatures*: the signatures labeling the nodes are derived from the binary code of the instructions by information compaction through a signature function S (e.g., a checksum, a Linear Feedback Shift Register (LFSR), etc.) as shown in Fig. *5-3*. The run-time signatures are derived by the watchdog concurrently, by compacting the instruction code captured on the bus.

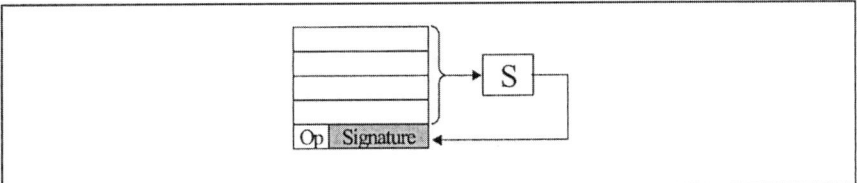

Figure 5-3. Derived signature basic technique.

2.1 Assigned run-time signature control-flow checking

Assigned run-time signature control-flow checking techniques label the nodes with signatures and check whether the run-time sequence of signatures is correct. Early methods checked the sequence by recording the execution and comparing it with previously determined sequences. In [145], some checkpoint in the sequence were assigned with distinct prime number signatures and compressed into a check symbol variable (through a repetitive multiplication). At the end of the run, the run-time labels are decomposed and checked for correctness.

2.1.1 Structural Integrity Checking (SIC)

This method was first proposed in [137]. The basic idea was that it is possible to parse the program to be executed by the main processor (which is supposed to be coded in a high-level language) and identify some common constructs (concatenation, selection, repetition and abstraction). In this way, nodes are identified and signatures (called *labels* in this case) are assigned to each of them. Once the code analysis and signature generation is concluded, the method is able to automatically generate

- a modified program for the main processor (named *Labeled Structured Program*) which differs from the original program only because it additionally includes special instructions devoted to send signatures to the watchdog
- a program for the watchdog, named *Structural Reference Program*. This program mimics the structure of the original program: it does not contain any computational part (as in the main processor program), but only the instructions for receiving signatures and checking for their correctness. The computational requirements for the watchdog are therefore much lower than for the main processor, making it much simpler and cheaper.

Interestingly, the string of signatures produced by the Labeled Structured Program can be described by a context-free grammar: the Structural Reference Program can be seen as implementing an automaton able to

accept exactly those strings described by the context-free grammar, and to
reject any other string.

Fig. *5-4* shows an example program (coded in Pascal): Fig. *5-5* reports
the Labeled Structured Program and the Structural Reference Program
produced by the application of the method.

```
      .
      .
      .
read(input, number);
repeat
  if number < 0 then begin
    negsum := negsum + number;
    number := number * -1;
  end;
  sum := sum + number;
  read(input, number);
until (number = 0) or (eoln(input));
      .
      .
      .
```

Figure 5-4. Example program

The SIC method offers a good coverage of those faults causing a change
in the order of execution of the nodes. However, two types of errors escape
the detection capabilities of the method:

- faults causing the control flow to produce an order of execution of the
  nodes which is different than the correct one, but still compliant with the
  context-free grammar extracted from the original program
- faults modifying the instructions corresponding to a node.

The major advantages of this method are that it woks on the high-level
code, and that any simple processor can be used to implement the watchdog.

Other methods that further elaborated the idea presented in [137] were
proposed in [138] and [145].

```
.                                    .

.                                    .

.                                    .
send(50);                            if signature <> 50 then error;
read(input, number);
send(187);                           if signature <> 187 then error;
Begin                                    Begin
repeat
        send(-82);                   if signature <> -82 then error;
        if number < 0 then begin       repeat
                                     (* loop terminated when signature
                                     different than -82 *)
            send(-12);                   if signature = -12 then begin
            negsum := negsum + number;   (* -12 means 'if' executed *)
            send(28);                        if signature <> 28 then error;
            number := number * -1;
        End else send(-13);          end;
        send(155);                   if signature <> 155 then error;
        Sum := sum + number;
        send(48);                    if signature <> 48 then error;
read(input, number);
until (number = 0) or (eoln(input)); until signature <> -82;
send(-83);
end;                                 end;

.                                    .

.                                    .
```

Figure 5-5. Labeled Structured Program (left) and Structural Reference Program (right).

2.2   Derived run-time signature control-flow checking

A derived run-time signature is a value assigned to each node. The term *derived* means that the signature is not an arbitrarily assigned value but calculated from the block's instructions. Derived signatures are usually obtained applying an *exor* function among the instruction opcodes or using such opcodes to feed a Linear Feedback Shift Register (LFSR). These values are computed at compile time and used as reference by the watchdog to verify the correctness of the executed instructions.

2.2.1   Embedded Signature Monitoring

Signatures are pre-computed by the compiler and generally stored within the application program. The watchdog processor compacts the instructions executed by the checked processor, and periodically compares the intermediate results of the compaction, which are the signatures, to pre-computed references. Many methods and approaches have been proposed in the literature. They differ in the definition of the node and in the representation of the reference information. Different processor architectures (traditional CISC or pipelined RISC) and configurations (mono or multiprocessors) are targeted, and trade-offs between the error coverage and the overhead introduced by the checking are taken into consideration.

The first method that introduced the Embedded Signature Monitoring technique was called BPSA (*Basic Path Signature Analysis*) [121], where a node is defined as a branch-free sequence of assembly level instructions and the reference signature is inserted at the beginning of each node. Two tag bits are used to differentiate signatures from other instructions in the node. The watchdog processor monitors the instruction bus of the processor and captures the reference signatures, using tag bits to differentiate them from the normal instructions. The checked processor executes a NOP instruction whenever a signature is fetched; on the other hand when a normal instruction is fetched, the watchdog processor computes the run-time signature, concurrently. A second tag bit signals the end of the node; the run-time signature is then compared to the reference one. A difference allows signaling the occurrence of an error.

The insertion of the embedded reference signatures increases the memory overhead of the checked program and reduces the performance. A proper compromise can be found between the error detection latency (the number of instructions involved in a node) and the overhead.

An improved method, called GPSA (*Generalized Path Signature Analysis*) [121] reduces the total number of signatures by checking sequences of nodes (called *paths*) rather than single nodes. Path sets (i.e., sets of possible paths starting from the same node) are defined and one signature is derived for each path set. The signature identifies the path set and so the computation of the run-time signature of each possible path in a set must be the same. This is obtained introducing some auxiliary signatures, called *justifying signatures* in some paths. These signatures are involved in the computation of the run-time signature, so that the same signature results at the end of each path.

In order to explain the approach in deeper details, let consider the control flow graph shown in Fig. *5-6*. The path set consists of four different paths P1 = $(V_1, V_2, V_4, V_5, V_7)$, P2 = $(V_1, V_2, V_4, V_6, V_7)$, P3 = $(V_1, V_3, V_4, V_5, V_7)$ and P4 = $(V_1, V_3, V_4, V_6, V_7)$. To each node V_i a signature $h(V_i)$ is assigned. In order to have a single signature computed for all the possible paths, justifying signatures are added at the nodes V_3 and V_6, according to the following formula:

$$h'(V_3) = h(V_3) \oplus h(V_2)$$

$h'(V_6) = h(V_6) \oplus h(V_5)$

After this modification, the signatures of the paths are the following:

P1: $H1 = h(V_1) \oplus h(V_2) \oplus h(V_4) \oplus h(V_5) \oplus h(V_7)$

P2: $H2 = h(V_1) \oplus h(V_3) \oplus h'(V_3) \oplus h(V_4) \oplus h(V_5) \oplus h(V_7) = H1$

P3: $H3 = h(V_1) \oplus h(V_2) \oplus h(V_4) \oplus h(V_6) \oplus h'(V_6) \oplus h(V_7) = H1$

P4: $H4 = h(V_1) \oplus h(V_3) \oplus h'(V_3) \oplus h(V_4) \oplus h(V_6) \oplus h'(V_6) \oplus h(V_7) = H1$

The common signature of the path set composed of these four paths is H1 and is stored at the node V_1.

The total number of stored signatures is three, which is less than the seven signatures stored following the *BPSA* method [121].

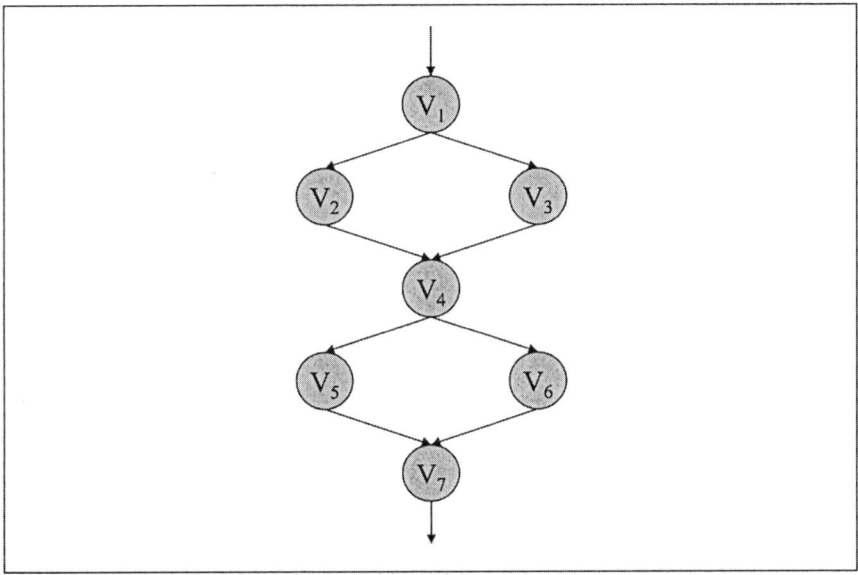

Figure 5-6. An example of control flow graph.

Wilken proposed in [118][119] an optimized method with the goal to find a set of justifying signature locations that causes minimum overhead. Wilken demonstrated that justifying signatures placed on certain nodes or arcs of the control flow graph cause less performance and/or memory overhead than at other locations. Using representative inputs, a program's execution profile can identify nodes and arcs that are visited infrequently, where justifying signature placement causes lower performance overhead. Moving from this analysis, each program graph node or arc i is labeled with cost c_i, which is a function of the performance overhead p_i and memory overhead m_i for placing a justifying signature there:

$$c_i = k_p p_i + k_m m_i$$

where k_p and k_m are non negative constants that are used to find the best trade-off between p_i and m_i.

The algorithm defined by Wilken to find the best placement starts from the transformation of the program control flow graph. A node X is added to the program graph, and all program exit arcs are connected so they are incoming to node X. This modified program is then transformed into a weighted *undirected graph*. Nodes and arcs in the modified program graph are modified according to the following rules:

- Node i is represented in the undirected graph by vertices v_{i1} and v_{i2}, joined by an undirected edge e_i
- Arc j is represented in the undirected graph by an undirected edge e_j; an arc j that is outgoing from node l and is incoming to node k, edge e_j connects vertices v_{l2} and v_{k1}.

For each node or arc i in the program graph the cost c_i is assigned to the corresponding edge e_i. The edge corresponding to the node X is assigned infinite cost in order to make impossible that a signature will be placed there. Fig. 5-7(a) shows the modified program graph and Fig. 5-7(b) shows the correspondent transformation into an undirected graph.

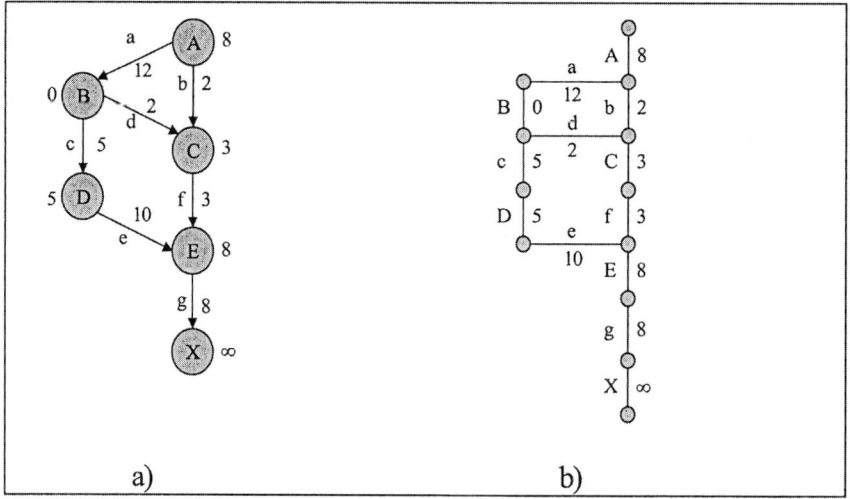

a)   b)

Figure 5-7. Program control flow graph transformation. a) Modified program graph. b) Undirected graph.

The proposed algorithm finds the cycle-free spanning tree for which a unique path exists between two nodes. The justifying signature with minimum cost is obtained by finding a minimum cost deleted edge set, and by placing justifying signatures on the corresponding nodes and arcs. The complement of a minimum cost deleted edge set is a non-deleted edge set that forms a maximum spanning tree, for whose computation several efficient algorithms exist. For the undirected graph shown in Fig. *5-7* the edge set *{A, C, D, E, X, a, b, c, e, f, g}* forms a maximum spanning tree. The complement set is *{B, d}*, and justifying signatures are placed on the corresponding node and arc at the optimal cost of 2.

CSM (*Continuous Signature Monitoring*) [116-117] presents an evolution of the GPSA approach. The control-flow error detection is improved by inserting a random distribution of intermediate signatures. The number of signatures is limited by a lower bound. The program is partitioned into the minimum number of paths, and one justifying signature instruction is added to each path. Using this method CSM is shown to reduce the number of signatures by as much as 3 times. CSM presents some novelties to reduce the latency. The previously proposed signature monitoring techniques encode an instruction sequence by embedding signatures in the vertical direction (as shown in Fig. *5-3*). Error detection latency can be high using this strategy because detection is delayed until the signature is checked at the path's end: to reduce the latency a signature monitoring approach using a horizontal strategy is proposed. Fig. *5-8* shows the *h* bits added horizontally to each word for storing a horizontal reference signature. The function *H* generates the horizontal signature for word *j* by operating on the instruction sequence from the path's beginning through word *j*. Horizontal signatures reduce detection latency because the monitor checks a signature at each program location; moreover, they cause no performance loss, because the signatures are fetched in parallel with the program code. However, they provide lower error detection coverage than vertical signatures for constant memory overhead.

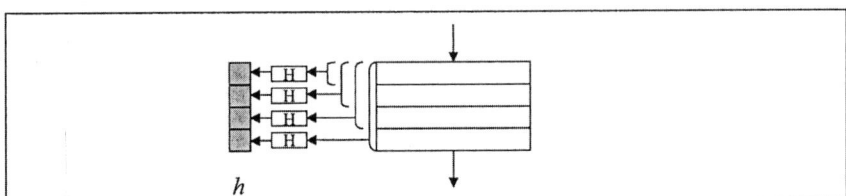

Figure 5-8. Horizontal signatures.

CSM proposed to combine horizontal and vertical signatures so that a short error-detection latency is ensured by the horizontal signatures, while error detection coverage is provided by the vertical signature. Fig. *5-9* shows a path encoded with signatures in two dimensions (horizontal and vertical). The signature compiler first generates the vertical reference signature using the function *V*, and then generates a horizontal reference signature for each location (including the vertical signature instruction) using the function H. During execution, the watchdog regenerates both run-time signatures, and compares them with their respective reference signatures. An experimental evaluation of the CSM approach carried about by the authors shows an estimated memory overhead ranging from 4 to 11%. In order to evaluate the fault detection capability, errors affecting the control flow have been considered. A control flow error modifies the flow from a correct location to a different one. The authors analyzed the control-flow error detection, i.e., the capability of the method to detect an erroneous flow, which is estimated to 99.9999% for a 32 bit processor.

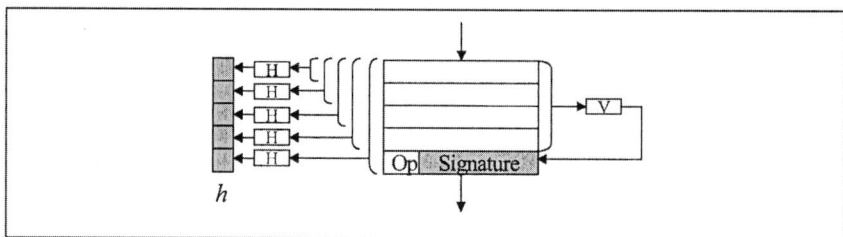

Figure 5-9. Combining vertical and horizontal signatures.

The approach proposed by Upadhyaya and Ramamurthy [130] considers a different approach, based on the so-called *tag instructions*: the signature for a sequential code is derived by applying a signature generation function successively on the opcode until the signature forms an *m-out-of-n* code for a specified *m* and *n*. A *n* bit code is an *m-out-of-n* code if and only if it has *m* 1's bits. The location in the memory that corresponds to an *m-out-of-n* code is tagged as a checkpoint for comparison. If the last instruction in the block does not form an *m-out-of-n* coded signature, a checkpoint must be force at the end of the block. Moreover, when a branch instruction is reached at the end of a block, the signature accumulation is continued along the branch, including the branch instruction opcode. A signature checkpoint is forced to check correct flow, this can be inserting an additional byte per branch and adjusting the accumulated signature to form an *m-out-of-n* code.

Tagged instruction are inserted at the compile time. During the execution phase, the generated signature at a tagged location is checked to determine whether it forms an *m-out-of-n* code. If it fails to form an *m-out-of-n* code at the tagged location, an error is signaled.

ISIS (*Interleaved Signature Instruction Stream*) has been first presented in [110]. The main idea is to include the signature expected for each block at the beginning of the block itself, in a code word which is not executed (nor fetched by the main processor); this code word is obtained by reorganizing the code and inserting these special words after branch instructions.

In this way the execution overhead introduced by the method is rather limited (it has been evaluated to be lower than 7% for a processor similar to the MIPS R3000 RISC processor). The memory overhead induced by the method for the same processor is in the range between 15% and 30%. Clearly, the watchdog module must work in close connection with the main processor and must be able to independently access the instruction cache (that must be transformed into a dual-port one), resulting in a non trivial implementation of the hardware part required by the method.

OSIRIS (Another Interleaved Signature Instruction Stream) is derived from ISIS and has been first proposed in [109]. To reduce the memory overhead resulting from signature embedding, bits in the instructions that are left unused are exploited to store signatures.

A watchdog module is inserted into the processor and concurrently performs the decoding of instructions, while the main processor performs the same operation. The watchdog is able to identify the signatures embedded in the instruction bits and to compare them with the signature of each block, that are computed on-line.

Clearly, this approach requires that a sufficient number of bits in the instruction coding are left unused (which can be the case in several RISC architecture). If these bits are not sufficient, NOP instructions are inserted to store the signatures.

As another limitation, this approach requires quite a deep intervention in the processor architecture to implement the watchdog module.

2.2.2 Stored reference

When a control flow checking mechanism belonging to this family is adopted, the reference information is stored in the local memory of the watchdog processor in some form (e.g., list, adjacency matrix). Each time a run-time signature is received or computed by the watchdog processor then the stored reference is searched to find out whether the signature is a valid successor of the previous one. To handle the stored reference, the implementation of the watchdog processor includes a general scarch and compare engine.

The main advantages of this approach are the following:

1. there is no performance overhead
2. the watchdog processor is independent from the monitored system
3. one watchdog can be shared among several processors under test.

A first method, called RMP (*Roving Monitoring Processor*) was presented in [129]. The method is able to detect faults in a multiprocessor system. It is based on a Hardware Signature Generation (HSG) for each

Application Processor (AP). The task of the HSG module is to generate a stream of signatures for its correspondent AP. The generated signatures are stored in a Signature Queue (SQ). The RMP processor stores a signature graph in its local memory. It samples the SQ memory, reading the signature generated by the APs and checks whether it belongs to a node which can be reached from the previous (checked) one. The system is able to check a multiprocessor system. Each signature word contains a signature and the processor ID. The whole architecture of the RMP method is shown in Fig. *5-10*.

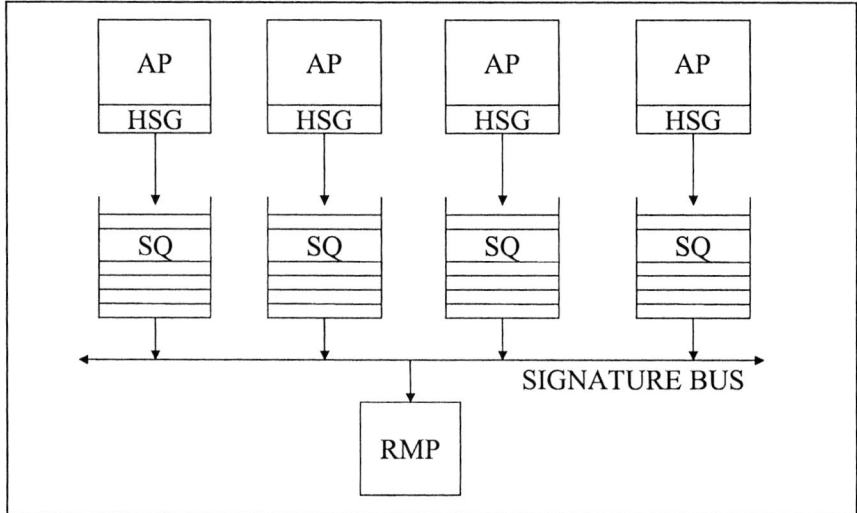

Figure 5-10. RMP monitoring system.

In the *Checker* approach [132] the signatures are generated by hardware generators attached to the application processors. The reference signatures are downloaded into the watchdog processor before the program run. Similarly to the RMP approach, the system is based on a Signature Generator (SG) added to each AP. The SG generates the program signatures at run-time and sends them to the watchdog.

The monitored program is divided in small sections, in such a way that there is only a small number N of signatures for every program section. For instance, for a 16 bit signature some possible values for N are 64, 128 or 256. Each program section includes as many sequences of contiguous instructions as required to contain N signatures. For each program section there is a correspondent segment in the watchdog processor memory where all the signatures of that program section are stored. Every time the SG module sends a signature to the watchdog, it also sends the address of the last instruction of the corresponding section: in this way the watchdog can identify the segment the signature belongs to. A run-time signature is considered correct if it is among the ones stored in this section. The verification is done using fast associative search. Fig. *5-11* shows the signature organization adopted by this approach.

The basic idea of the signature verification lies in the fact that the probability of a wrong signature being equal to any other signatures in the same section is very low, if a random distribution of the signatures is assumed.

The main advantages of this method are that the overhead of the reference memory is reduced, since the structural information is not stored and the associative search is fast enough to serve several processors in parallel.

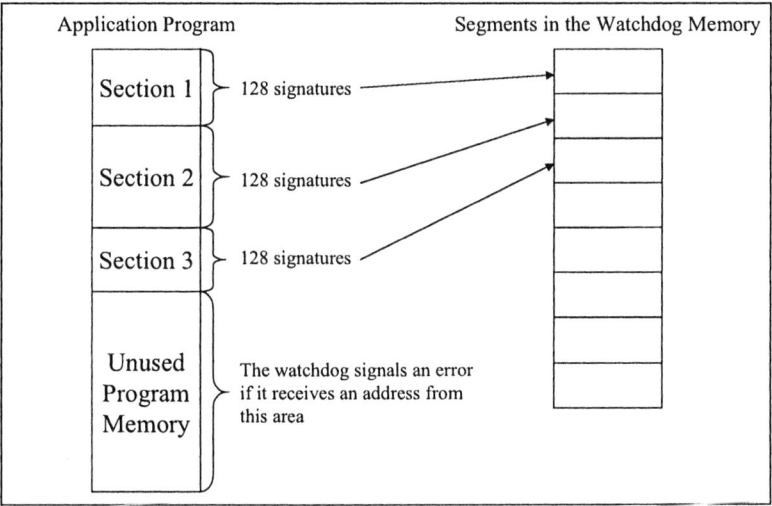

Figure 5-11. An example of program organization for the Checker approach.

2.2.3   Reference program

When this family of methods is considered, the watchdog processor is a specialized (seldom a general purpose) processor, which executes a reference program with the same control flow graph as the program under check. The architecture of the processor is simple, since most instructions are checking and control transfer instructions, and no ALU is needed

A first method proposing such a scheme was called *Cerberus-16* [122]. The basic idea behind the method is that the program executed by the checked processor has the same control flow graph of the program executed by the watchdog processor, which has a restricted instruction set.

The program executed by the checked processor is mapped in compilation time to a program of the watchdog processor which has the same control flow graph. Any program graph can be represented by a set of watchdog instructions. A suitable instruction is used to represent each node, depending on it. When the checked processor executes a branch (at the end of a node), this is signaled by a *branch detecting unit,* and the watchdog executes the instruction correspondent to that node. Each node in the program executed by the watchdog consists of a single instruction with the following format:

$$OP \; Z, \; [L], \; [D]$$

where Z, L, and D represent the node size, the next node address and the node signature, respectively.

After Z instructions have been executed by the checked processor, the watchdog waits a signal from the branch detecting unit to indicate whether a branch was taken by the main processor or not. If it detects a branch, then the watchdog executes a branch using the L field to compute the target address of the next instruction, otherwise the watchdog executes the next instruction in sequence. The reference signature of a path is computed by a *data compression unit* (implemented for example by an exor function). At the beginning of a checked path, the data compression unit is initialized, and at the end of the path the reference signature, stored in the D field, is compared to the run-time one. If the checked processor executes a branch (at the end of a node) then it is signaled by the *branch detecting circuit* and the watchdog takes the next instruction given in the code of the actual one (otherwise the successive instruction is executed). This way, each instruction of the watchdog checks a node of the main program.

A second approach, called *Watchdog Direct Processing*, has been proposed in [131]. The watchdog processor directly monitors the addresses of the main processor. The watchdog processor contains the information necessary to detect the nodes of the control flow graph. The signature is checked cach time a destination node is reached, which leads to a reduced detection latency. The propagation of the error in the system is therefore reduced and furthermore, the path followed through the control flow graph is completely identified, making easier the error recovery process. The watchdog program contains one instruction for each node in the application program. Each instruction includes three fields: the opcode (type of node), the address of the associated instruction in the application program, and the reference information.

On each node the following information are necessary:
1. type of node (initialization node, destination node, unconditional or conditional branch, unconditional or conditional branch to subroutine, return from subroutine)
2. address of the instruction associated with the node
3. signature reference value on this node
4. address of the destination node (if any).

The watchdog processor has two main tasks: it computes the signature of the executed instruction sequence and it detects the nodes reached by the main processor. When a node address is reached, the watchdog processor checks the current signature, it updates the next node address and verifies the destination address using the information available on the destination node. In case of mismatch it sets an error signal. The authors of the paper demonstrate that the total memory overhead introduced by the application watchdog program is lower than the overhead introduced by Cerberus-16 and the complexity of the implemented watchdog is acceptable, allowing to integrate such a watchdog in a single chip.

3.   MEMORY ACCESS CHECKING

Faults can affect the correctness of the memory references. Classical methods adopted to detect illegal accesses to the memory system may operate at the operating system level (protection violation) or at the circuit level (functional testing). In any case, there are several categories of errors that may not be detected by them. A watchdog can be inserted to verify the correctness of the memory accesses.

The method proposed by Namjoo and McCluskey [123] defines an *object* as the code and data address ranges in the memory. This method does not present any alteration of the application source code, and is based just on the capability of the watchdog processor to verify the validity of the memory references.

Each object has a capability (*access right*) for accessing another object (e.g., read, write, execute). The set of objects and the corresponding capabilities are represented by a directed graph (*object graph*), where a vertex represents an object and the edge shows the access right of an object to another one. The object graph is stored in the watchdog processor in a table called *Segment Access Table* (*SAT*). The *SAT* has one row for each code object; the entries in each row are the access rights given to that object. A second table, called *Segment Map Table* (*SMT*), is stored into the watchdog processor and is used for mapping the physical page frame numbers (derived from the memory address bus) onto segment numbers representing objects. For each memory reference, the physical address is translated to a segment identifier (*segment ID*) using the *SMT*. The segment ID is used in turn as the address for accessing the *SAT*. Two segment IDs are required to access the *SAT*: the first is the segment ID (S_i) of the current code object (O_i); the second is the segment ID (S_j) of the object (O_j) referenced by the current object. S_i and S_j are determined from the physical address in each reference through the *SMT* table. An entry $SAT(S_i, S_j)$ in the *SAT* table shows the access right of the object S_i to the object S_j.

The watchdog monitors the physical memory accesses of the processor under test and derives the active object (from the address of the fetched instruction through an access to the *SMT* table), the accessed object (from the destination address of the instruction through an access to the *SMT* table), and the actual operation (analyzing the instruction itself). The access right required to perform the access is compared to the enabled access right stored in the object table and read out by the *SAT* at the entry $SAT(S_i, S_j)$; if there is no match a fault is detected and an access error is signaled. Fig. *5-12* shows a representation of the memory accessing checking executed by the watchdog processor. In Fig. *5-12* register Rx holds the segment ID of the current code segment (S_i) which is determined from the current memory reference using the mapping data in the SMT. The segment ID for the next reference to the memory (S_j) is also determined and loaded into register Ry by the watchdog processor. The entry $SAT(S_i, S_j)$ is read out from the *SAT* and is compared with the access requested by the CPU.

With this method the watchdog processor checks the validity of each access in parallel with the CPU operation. This operation is repeated for each memory reference.

This method guarantees no degradation in the system performance, since the checking is done in parallel with the main processor and any modification to the software is required. The limited complexity of the watchdog processor allows guaranteeing a low-cost solution.

The main disadvantage of this technique is that it is not able to cover all the possible faults. The main class of undetected faults are the ones that cause an incorrect operation (i.e., an operation that is incorrect under certain conditions, but may be correct in other ones). Incorrect operations cannot be recognized properly since not all the necessary information are embedded into the watchdog processor.

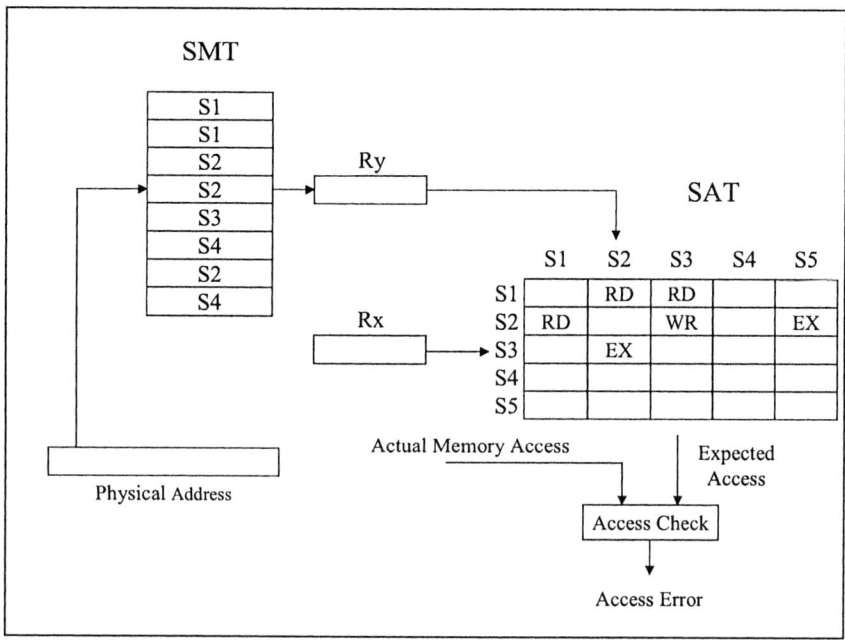

Figure 5-12. Memory Access Checking using a watchdog processor

4. REASONABLENESS CHECKING

A watchdog processor can be used to check the correctness of the manipulated data. Data errors can be detected by having the watchdog execute assertions concurrently. An assertion is an invariant relationship between the variables of a program. The assertions are inserted by the programmer at different points of the program, stating what he intends to be true for the variables. Assertions can be written on the basis of the specifications or of some property of the algorithm. They are usually based on the inverse of the problem, on the range of values that variables can assume, or the relationships between variables. The insertion of executable assertions within a program is described in Section 2.4, whereas the application of watchdog processors for the concurrent execution of assertions is summarized here.

The main objectives when devising such a kind of watchdogs is to keep their complexity as low as possible and to transfer the data from the main processor to the watchdog without any significant overhead. There are two alternatives to solve these problems: one for special purpose architectures and the other for general purpose one. In both schemes, the code of the assertions is stored into the local memory of the watchdog as a library of functions, and only the identifier of the assertion function and the data have to be transferred to the watchdog.

4.1   Watchdog methods for Special Purpose applications

In special purpose architectures the flow of data (e.g., the sequences of data values on the data bus) is often known and invariant. The watchdog can be designed to suit a particular application.

The solution proposed by Mahmood et al. [124] is based on data bus monitoring to recognize instructions that modify critical data. The code for the assertions is stored in the local memory of the watchdog and the instructions, which assign values to the variables, are tagged. The watchdog is able to capture the data by monitoring the data bus of the checked processor and capturing the tagged data. This approach has been adopted to problems that solve systems of equations using Gaussian elimination, discrete Fourier transform, eigenvalues, etc.

A different strategy can be used if the application is cyclic and uses a large number of global variables. This is the case of many real-time applications, i.e., telephone switching systems and digital flight control systems. The executable assertions that check the correctness of the values are stored in the local memory of the watchdog. Critical data, stored in global variables, are transferred to the watchdog by simultaneously writing to both the main memory and the local memory of the watchdog.

Cyclic applications are based on a repetitive data elaboration (e.g., data stored in global variables are processed with a predefined frequency, repeating a cycle of instructions).

Thanks to this property, the watchdog may exploit a dual buffer scheme to execute assertions. The first buffer is used to store the data captured by the watchdog and the second buffer is used to execute the assertions. At the end of each cycle data are moved from the first to the second buffer. Assertions are thus executed on the data captured during the previous cycle with a limited and acceptable latency.

4.2   Watchdog methods for General Purpose applications

In a general purpose architecture the watchdog cannot be previously designed and programmed according to a specific application. The main difficulty is then the transfer of data from the processor under test to the watchdog. The solution proposed in different papers ([125][126]) is based on message passing: the main processor writes into a shared buffer and the watchdog reads from it. Besides the shared buffer both the processors also have their local memories. The software structure is the following:

- Before execution, the program is modified by replacing the assertion functions with a single statement which transfers the data values and the identifier of the assertion function to the watchdog. The write statement can be the following:

  `write_buffer(assertion_code,space_needed,data)`

  where `assertion_code` is the assertion function identifier, `space_needed` is the memory space needed by the data, and `data` are the values of all the variables which are used in executing the assertion.

- Additionally, the code of the assertion functions is downloaded into the local memory of the watchdog processor.

- At run-time, the main processor writes to the shared buffer and the watchdog reads from it and executes the required function. If the logical result computed by the assertion function is false, then an error is signaled.

An example of the transformed programs for the main processor and the watchdog are shown in Fig. *5-13*.

```
        .                          main(
write_buffer(1, space_needed, data);
        .                          {
                                   Read_next(assertion_number);
        .                          Switch(assertion_number) {
        .                          case 1:  get(data);
write_buffer(n, space_needed, data);        assertion_1();
        .                                   Break;
        .                          case 2:  get(data);
        .                                   Assertion_2();
                                            Break;

                                       .

                                       .

                                   Case n:  get(data);
                                            Assertion_N();

                                   }
```

Figure 5-13. Main processor program (left) and Watchdog Program (right).

5.   COMBINED TECHNIQUES

Techniques belonging to this category aim at covering both faults causing changes in the control flow execution, and faults affecting the data.

5.1   Duplication and watchdog

A basic approach adopted to design dependable systems is to use redundancy: the simplest approach is based on duplication with comparison, where two synchronized processors execute a single application concurrently and an external comparator compares their outputs. When a mismatch appears, the comparator signals the occurrence of an error. The growth in computer microprocessor functionality increases the bus complexity, the working frequency and the number of processor pins, which makes the external comparison of the pins very difficult. This has encouraged designers to move the comparison mechanism into the processor. This feature is called Master/Checker (M/C) and is supported by many modern processors (e.g., those of the Pentium family, AMD K5 and MIPS R4000). The M/C architecture [160] is based on the duplication of processors: one processor operates in the Master Mode, and the other one in the Checker mode. Both processors run the same program and process the same data stream, fully clock synchronous. In the Intel Pentium family, such a duplication structure can be set without external components, as the necessary logic, called Functional Redundancy Checking (FRC), is integrated inside the chip. The processor configured as Master operates according to the bus protocol. The outputs of the Checker processor are tri-stated, so the outputs of the Master processor can be sampled. If the sampled values differ from the values computed internally by the Checker processor, the Checker signals this error. The M/C architecture has been experimentally evaluated in [160] showing that software faults are effectively detected, but CPU crashes occurred in the Master processor are not detected by the Checker processor.

A novel fault tolerant embedded architecture has been proposed in [161], based on two processors with Master/Checker (M/C) configuration and an external watchdog processor. The watchdog processor is constituted by a Triple Modular Redundancy of watchdog processors (TMR-WDPs) as shown in Fig. *5-14*. The TMR-WDPs contains three watchdog processors (WDPs) working in parallel, and comparing their outputs with a majority voting logic. The WDP must be able to recognize special I/O cycles and get run-time information from the workload program. The fault tolerant system exploits a Performance Monitoring feature available in modern superscalar processors. The Performance Monitoring feature includes some special internal counters, which can be configured to count the occurrences of processor events, such as cache hits, instruction committed and branches taken. The fault tolerant system proposed in [161] exploits the Performance Monitoring facilities available with the Intel Pentium family processors. Several internal registers, called *Model Specific Registers* (MSRs), reflect the internal processor status. Two of these registers, CTR0 and CTR1, are used as counters. These registers are programmed through another MSR register, called CESR. To access to the MSRs register, the processor provides two instructions WRMSR and RDMSR. An increment of the counters can also be reported on two external pins of the processor, called PM0 and PM1.

The error detection mechanisms, called *Committed Instructions Counting* (CIC) [159] are the following:

- The number of executed instructions in a basic block is internally counted; at the beginning of the basic block, an internal counter is set to zero and the pre-computed number of instructions of the basic block embedded in the program is sent out to the watchdog processor. During the execution of the basic block, the content of the internal counter is decreased by one each time an instruction is executed. At the end of the basic block, the content of the counter (computed at run-time) is also sent to the watchdog processor. The watchdog checks the two values (pre-computed and run-time) and signals the occurrence of an error if any discrepancy exists.

- The number of executed instructions in a basic block is externally counted by the watchdog through the PM0 and PM1 pins. At the entry point of the basic block the pre-computed number of instructions existing in the basic block is sent to the watchdog processor; the pins PM0 and PM1 operate as clock signals to increment a counter internal to the watchdog each time an instruction is executed; the watchdog signals an error if the run-time number of executed instructions differs from the pre-computed number.

- A unique signature is assigned to each basic block and it is sent to the watchdog at the beginning and at the end of that basic block. The watchdog signals an error if these two indices are different.

- A workload timer is located in the watchdog checking the upper limit on the workload execution time.

Experimental results demonstrated that the system composed of a M/C board and a TMR-WDP module is able to detect 100% of CPU crashes on the Master and on the Checker processors, and a percentage ranging from 91% to 95% of faults on the main memory.

The method is relatively weak in detecting faults modifying the transitions from one block to another.

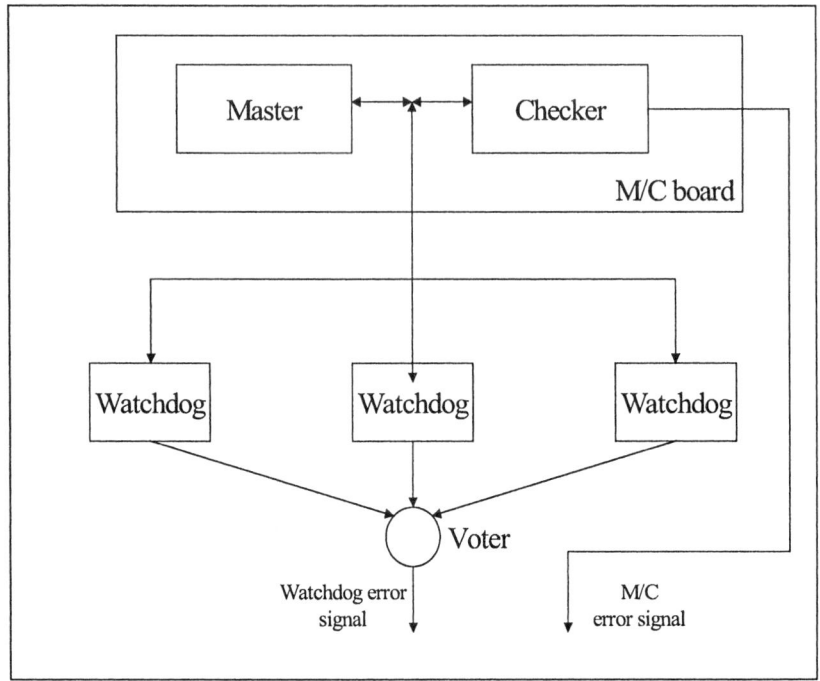

Figure 5-14. M/C Architecture and Watchdog processors system.

5.2   Infrastructure-IP

The technique proposed in [158] mainly addresses the fault tolerance properties of processor-based systems implemented on a single chip (also called *Systems on Chip*, or SoCs). When hardening SoCs, it is common not to be in the position of modifying the modules (also called Intellectual Property cores, or IP cores) implementing the processors, while some circuitry can be rather easily added outside the functional cores. The method integrates the ideas originally introduced in [157] and [156], where the two different issues of hardening the system with respect to control flow errors and data errors were separately faced, respectively.

In order to overcome the limits of the purely software approach presented in the previous chapters a hybrid solution tailored to be applied in SoC devices was proposed. The main idea is to adopt the approach described in [154] and [155], but to resort to a watchdog to reduce its cost and enhance its performance in terms of fault detection capabilities. When dealing with SoCs, the watchdog can be implemented as an additional module implemented on the same device, and corresponding to a so called *Infrastructure IP* (or I-IP[4]).

The result is a *hybrid* approach where fault detection–oriented features are still implemented in software, but most of the computational efforts are demanded to external hardware. In practical terms, the executed program allows the processor to communicate with an external circuitry through the SoC bus: by computing the received information, this circuitry determines incorrect executions.

The goal is to devise a method that can be easily adopted in the typical SoC design flow; this means that the method requires minimal changes in the hardware (apart from the insertion of the I-IP), while the software is simplified with respect to the purely software fault detection approach proposed in [154] and [155]. A further constraint is the flexibility of the approach: any change in the application should result in software changes, only, while the hardware (including the I-IP) should not be affected. This means that the I-IP does not include any information about the application code, but is general enough to be able to protect any code, provided that it has been hardened according to the suggested approach.

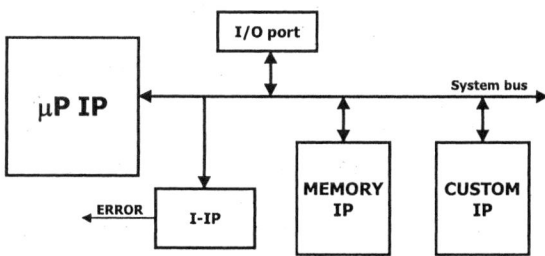

Figure 5-15. Architecture of the generic SoC system including the fault detection-oriented I-IP.

The proposed I-IP is connected to the system bus as an I/O peripheral interface. This means that the I-IP can observe all the operations performed on the bus by the processor, and can be the target for some write operations performed by the processor at specific addresses of the memory or I/O

---

[4] An Infrastructure IP is defined as an IP core deprived of any purely functional role, but introduced in the SoC to support ancillary features, such as debug, test, or reliability (as in the case we are presenting).

address space (depending on the adopted I/O scheme). When the I-IP detects an error, it activates an ERROR signal, which can be sent either to the processor, or to the outside, depending on the preferred recovery scheme. The architecture of the system including the I-IP is reported in Figure *5-15*.

The method can be introduced more easily by first considering in a separate manner the techniques adopted for dealing with faults affecting the code and those dealing with faults affecting the data. However, the two sets of techniques are supported in an integrated manner by the I-IP, resulting in even higher fault detection capabilities with respect to the purely software approach.

5.2.1   Support for Control Flow Checking

The basic idea behind the proposed approach for checking the correct control-flow execution (presented in [157]) is that we can simplify the hardened code and improve its performance by moving in hardware the control flow checks. According to the proposed solution, the code is in charge of signaling the I-IP when a new basic block is entered. Since the I-IP is not intended to record any information about the application code, the hardened program must send to the I-IP all the information required to check whether the new block can be legally entered given the list of previous blocks. The I-IP records in an internal register the current signature. Once it is informed that a new block is entered and it has received the list of blocks that can reach legally the new block, it checks whether the stored signature is included in this list. If not, the ERROR signal is raised. Otherwise, the current signature is updated with the signature of the new block.

In order to support the communication between the processor and the I-IP, two high-level functions are introduced, named `I-IPtest()` and `I-IPset()`. Their role is the following:

- `I-IPset(B`$_i$`)` informs the I-IP that the program has just entered into basic block B$_i$.
- `I-IPtest(B`$_j$`)` informs the I-IP that block B$_j$ belongs to the set of the predecessors of the newly entered block.

The I-IP contains two registers A and B that can be accessed by the processor by performing a write operation at a couple of given addresses X$_A$ and X$_B$. The two functions `I-IPtest()` and `I-IPset()` are translated into write operations at the addresses X$_A$ and X$_B$, respectively, thus resulting in a very limited cost in terms of execution time and code size. The parameter of each function is written in the register, thus becoming available to the I-IP for processing. A sequence of calls to the two functions should be inserted in the code at the beginning and at the end of each block B$_k$. First, a call to `I-IPtest(B`$_i$`)` is inserted for any block B$_i$ $\in$ prev(B$_k$). Then, a call to `I-`

`IPset(B`$_k$`)` is inserted. When noticing a write operation on register A, the I-IP set or reset an internal flag depending on the result of the comparison between the function parameter and the internally stored signature. When noticing a write operation on the register B, the I-IP verifies the value of the flag and possibly activates the ERROR signal. Otherwise, the signature of the current block is updated using the value written in the register B.

```
B1:  x = 1;
     y = 5;
     i = 0;
B2:  while( i < 5 ) {
B3:     z = x+i*y;
        i = i+1;
     }
B4:  i = 2*z;
```

Figure 5-16. Example program fragment.

```
I-IPtest(S₀,₂);
I-IPset(S₁,₁);
x = 1;
y = 5;
i = 0;
I-IPtest(S₁,₁);
I-IPset(S₁,₂);
while( i < 5 ) {
   I-IPtest(S₁,₂);
   I-IPtest(S₃,₂);
   I-IPset(S₃,₁);
   z = x+i*y;
   i = i+1;
   I-IPtest(S₃,₁);
   I-IPset(S₃,₂);
}
I-IPtest(S₁,₂);
I-IPtest(S₃,₂);
I-IPset(S₄,₁);
i = 2*z;
I-IPtest(S₄,₁);
I-IPset(S₄,₂);
```

Figure 5-17. Control-flow check according to the hybrid approach.

```
I-IPtest(S0,2);
I-IPset(S1,1);
x0 = 1; x1 = 1;
y0 = 5; y1 = 5;
i0 = 0; i1 = 0;
I-IPtest(S1,1);
I-IPset(S1,2);
while( i0 < 5 ) {
    I-IPtest(S1,2);
    I-IPtest(S3,2);
    I-IPset(S3,1);
    z0 = x0+i0*y0; z1 = x1+i1*y1;
    i0 = i0+1; i1 = i1+1;
    I-IPtest(S3,1);
    I-IPset(S3,2);
}
I-IPtest(S1,2);
I-IPtest(S3,2);
I-IPset(S4,1);
i0 = 2*z0; i1 = 2*z1;
I-IPtest(S4,1);
I-IPset(S4,2);
```

Figure 5-18. The full implementation of the hybrid approach.

A code portion for the example introduced in Fig. *5-16* that adopts the proposed approach, i.e., sending information to the I-IP, is reported in Fig. *5-17*.

Two functional parts can be distinguished in the I-IP to execute concurrent control-flow checking: Bus Interface Logic, and Control Flow Consistency Check Logic. Such schematic circuitry subdivision is highlighted in Fig. *5-20*.

The Bus Interface Logic implements the interface needed for communicating with the processor bus.

The Control Flow Consistency Check Logic is in charge of verifying whether any control flow error affects the application expected behavior, and to inform the system through the error signal if error detection happened. The I-IP is internally provided with both the circuitry to store and update the current signature each time data are sent from the processor: such circuitry calculates at run-time the value of the masks according to the technique proposed in [155].

5.2.2   Support for Data Checking

When considering the faults affecting the data, the approach is based on the idea of moving in hardware (i.e., charging the I-IP of) the task of comparing the two replicas of a variable each time it is accessed for read purposes. In this way the hardened code is significantly simplified: not only its size is reduced and the performance increased, but a number of conditional jump instructions are removed, thus reducing the risk for additional faults affecting the code.

To implement the above idea, the I-IP must monitor the bus, looking for memory read cycles. In principle, the I-IP should simply identify the two cycles accessing the two replicas of the same variable, checking whether their value is identical. If not, an error is detected.

In practice, implementing this idea requires a mechanism allowing the I-IP to know the addresses of the two replicas of the same original variable and to understand whether a given address corresponds to the first or second replica. A solution to this issue will be described further in this section.

Moreover, it is important to note that the two bus cycles accessing to the two replicas of the same variable are not necessarily consecutive. In fact, the compiler often reorganizes the assembly code so that instructions are re-ordered in such a way that the two instructions are interleaved with others. However, in developing the I-IP the authors assumed that the compiler never modifies the code in such a way that the second replica of a variable is accessed before the first replica. To tackle this issue, the I-IP contains a CAM memory, which is used to store the address-data couple corresponding to each variable accessed in memory, whose replica has not been accessed, yet. The CAM is indexed with the address field. More in details, the I-IP implements the following algorithm:

- If a memory read is detected on the bus, the address and data values are captured.
- If the read operation relates to the first replica of a variable, a new entry is inserted in the CAM, containing the just captured address and data values.
- If the read operation relates to the second replica of a variable, an access is made to the CAM:
  - If an entry with the same address is not found, the ERROR signal is raised.
  - Otherwise, the data is compared with that stored in the CAM entry and the ERROR signal is raised in the case of a mismatch.
  - The entry is removed from the CAM.

This simple algorithm has several interesting properties. It detects all the faults affecting the data that can be detected by the purely software approach. It can be straightforwardly (and inexpensively) extended to deal with write operations, too. A separate CAM is reserved for entries related to write operations. Thanks to this extension, some faults that cannot be detected by the corresponding purely software approach are detected by the hybrid one. When the end of a basic block is reached, the CAM should be empty, since the two replicas of all the variables should have been accessed. If this is not the case, an error (likely, a control flow error) has happened: the ERROR signal is raised.

An example of how the hardened code of the example should be modified according to the above approach is reported in Fig. *5-18*.

As we mentioned before, an efficient mechanism is required to allow the I-IP to understand whether a given address identifies the first or second replica of a variable, and to compute the address of the first replica once that of the second is available. The solution proposed in [156] assumes that the data segment of the program is divided in two portions, as shown in Figure *5-19*. The upper portion contains the first replica of each variable, while the lower one stores the second replica. This solution can be easily implemented acting on the options of C compilers.

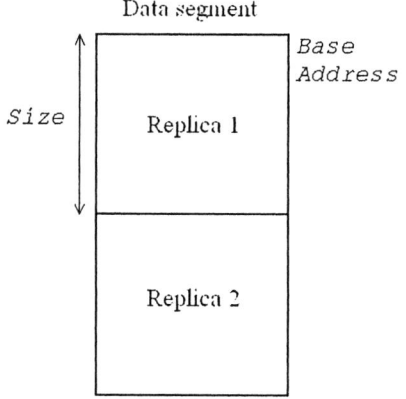

Figure 5-19. Data segment of the hardened program.

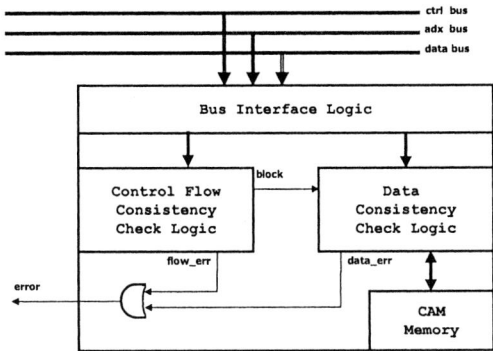

Figure 5-20. Schematic architecture of the I-IP implementing Control Flow and Data Checking.

The above assumption about variable location in memory easies the task of dealing with the two replicas of the same variable. More in details, as soon as a memory access cycle is detected on the bus, the two fields (adx, data) are extracted, corresponding to the address and value of the accessed variable, respectively. Being **Base** the beginning address of the data segment and Size the size of each portion of the segment, if adx < **Base + Size**, then the first replica of the variable is currently being accessed; otherwise, the second replica is being accessed. To compute the address of the first replica when the address adx_2 of the second is available, the following expression is used:

$$adx_1 = adx_2 - Size$$

Three functional parts can be distinguished in the I-IP circuitry devoted to execute concurrent data checking, as reported in Fig. *5-20*: these parts are named Bus Interface Logic, Data Consistency Check Logic, and CAM Memory.

The Bus Interface Logic is shared with the circuitry devoted to control flow checking and implements the interface for accessing to the processor bus. It is able to decode the bus cycles being executed and in case of read or write cycles to the memory, it samples the address (adx) and the value (data) on the bus. Sampled addresses and values are then forwarded to the Data Consistency Check Logic.

The Data Consistency Check Logic implements the consistency checks verifying whether any data stored in the memory or the processor has been modified. For this purpose, as soon as a new couple (adx, data) is extracted from the bus, it computes the address of the corresponding replica, accesses to the CAM memory, and verifies whether the searched entry exists. In the positive case, it compares data with the data field of the entry (possibly raising the error signal in case of mismatch) and then removes the entry from the CAM. In the negative case, it inserts a new entry in the CAM.

Moreover, considering the program structure presented at the beginning of this paragraph, each instruction into a basic block has a replica within the same block. Consequently, we can assume that the CAM memory is empty when a new basic block is entered. To cope with this assumption, the Data Consistency Check Logic receives a block signal, generated by the Control Flow Consistency Check Logic. Such signal is asserted when exiting a basic block: the content of the CAM Memory is then checked and, if not empty, the error signal set on.

The proposed I-IP design, whose schematic is shown in Fig. *5-19*, can be easily adapted to different processors: both the CAM Memory and Consistency Check Logic modules are parametric and can thus be reused for different address and data sizes. Only the Bus Interface Logic needs to be reworked for adapting to the bus protocol implemented by different processors. When the I-IP is introduced in a SoC, the only customization required concerns the addresses X_A and X_B of the two registers written by the **I-IPtest()** and **I-IPset()** procedures, respectively, and the values of Base and Size.

5.2.3   Error detection capabilities

To experimentally assess the effectiveness of their hybrid approach, the authors of [158] developed first a categorization of the possible faults affecting the memory elements of a system, and then theoretically analyzed the error detection capabilities of the proposed method. Faults can be divided in the following types, according to the module affected by the considered bit flip: memory code area, memory data area, and processor internal memory elements.

When considering the first fault type, the processor instruction set can be seen as divided into two instructions categories: functional instructions (executing some sort of processing on data, such as transfer, arithmetic operations, bit manipulation, etc.), and branch instructions. Consequently, whereas the modified bit in the code area belongs to an opcode, the following categories can be introduced:

- functional_to_branch: the modified bit in the opcode transforms a functional instruction into a branch instruction.

- branch_to_functional: the modified bit in the opcode transforms a branch instruction into a functional instruction.
- functional_to_functional: the opcode of a functional instruction is transformed into another functional instruction:
  - With the same number of operands.
  - With a different number of operands.
- branch_to_branch: the opcode of a branch instruction is transformed into another branch instruction:
  - With the same number of operands.
  - With a different number of operands.

In the case of a functional_to_branch code modification, the program flow is guaranteed to change; if the target of the branch introduced by the fault is out of the basic block boundary, both software and hybrid detection mechanisms detect the fault. On the contrary, when the branch target is inside the currently executed basic block, software detection may fail, while hybrid successfully copes with most of such faulty behaviors, thanks to the additional check on the CAM memory emptiness performed at the end of each block. These faults could also lead to a timeout, if the target of the faulty jump is a previous instruction within the same basic block.

Faults belonging to the branch_to_functional category also cause a change in the program control flow. If the new instruction has the same number of operands than the original, the detection is guaranteed by both approaches, thanks to the consistency check (for the software approach) and to the data checking techniques (for the hybrid approach). On the other hand, if the new instruction has a different number of operands, in the software approach the fault may not be detected because the consistency checks can only evaluate the equivalence between two variables and are not able to evaluate possible misalignments into the code, while the hybrid approach is able to detect such kind of faults thanks to its capability to store all the memory accesses and verify possible unbalanced memory accesses.

For functional_to_functional code modifications, if the number of required operands of the exchanged instruction is the same in the original and faulty instructions, both approaches are able to detect the fault. Unfortunately, if the number of operands is changed, neither the software nor the hybrid approach can always guarantee the detection: in this case, it is possible that the modified program execution continues until the end, producing a wrong answer, even if the probability of this situation is really low. In fact, the program usually backs to its normal flow, with unexpected CAM memory content.

For branch_to_branch code modifications, in the case of unchanged number of operands, it is possible that the modified program execution continues until the end, producing a wrong answer due to incorrect condition evaluation, or more frequently, an endless loop finally resulting in the time-out condition is entered. If the number of required operands is modified, it can happen that the end of the program is reached and a wrong answer produced. As for the functional_to_functional code modification, the probability of such event is low as the program usually backs to its normal flow, with unexpected CAM memory content. In fact, the program usually backs to its normal flow, with unexpected CAM memory content.

If the faulty bit corresponds to the operand of an instruction, the following unexpected program behaviors have to be investigated:

- wrong_memory_access: the modified operand is the address of a variable.
- wrong_immediate_value: the modified operand is an immediate value.
- wrong_branch_offset: the modified operand is the target of a branch instruction.

Considering the faults belonging to the wrong_memory_access category, they are covered by all the approaches, as they modify only one of the two replicas; therefore, the fault is detected by the data checking techniques.

The wrong_immediate_value fault category is covered by both the software and hybrid approaches. The following cases must be considered: if the involved instruction is a comparison executed immediately before a branch instruction, the fault is covered by the data checking techniques at the beginning of the basic block, otherwise the fault effect modifies the value of one replica of the variables and is detected by the data checking techniques.

Finally, when a branch is made to a wrong address, that is the wrong_branch_offset code modification, a wrong answer is never produced for both the techniques analyzed; however, it is possible that such modification leads to the timeout condition.

Considering the faults affecting the data area, the effects of the faults can be classified as follows:

- wrong_elaboration: the value read from the memory is wrong.
- wrong_branch_condition: a branch condition is executed on a modified (and thus incorrect) value.

Both the software and the hybrid strategy guarantee the detection of a wrong_elaboration fault affecting the system thanks the data checking techniques, while, if a wrong_branch_condition fault occurs, we can distinguish between two situations: the variable is nevermore accessed during the program, so we have a wrong answer; the variable is accessed again and a mismatch with its replica is observed. To avoid the former case, a read operation of the second replica of the variable is inserted exactly at the beginning of the basic block following the branch.

Concerning the effects of a single fault affecting the content of the processor registers, the following cases should be considered:

- wrong_general_purpose_value: a general purpose register stores a wrong value.
- wrong_configuration_value: the processor is configured incorrectly.

Faults in the Wrong_general_purpose_value category are usually detected by both the software and hybrid approaches thanks to the data checking techniques; however, sometimes two transfer instructions can read the value of the same register, then copying it into two replicas. Such situation is usually generated when code optimization is used by the compiler. Finally, the impact of wrong_configuration_value faults on the program execution depends on the processor configuration and usually results in a wrong answer or, more easily, in a timeout condition with both the approaches.

5.2.4 Experimental Results

To assess the effectiveness of their approach, the authors of [158] developed a prototypical implementation of the I-IP and exploited it for hardening a SoC including an Intel 8051 controller. For this purpose, the Infrastructure IP they proposed was described in VHDL and connected with a soft-core implementing the Intel 8051. Some benchmark programs were used to assess the properties of the hybrid approach in terms of detection capabilities and cost (memory overhead, performance slow-down, silicon area required by the I-IP).

To model the effects of SEUs, the authors of [158] adopted the *transient single bit flip* fault model, which consists in the modification of the content of a single storage cell during program execution.

The fault-detection ability of the approach were separately investigated, considering:

- SEUs modifying the content of the code memory area
- SEUs affecting the data memory area
- SEUs affecting the microcontroller's internal memory elements.

The fault-injection tool adopted for the experiments allowed accessing all the memory elements the processor embeds, with a suitable time resolution [150].

In setting-up the fault injection experiments, a crucial factor is the selection of the number of faults to inject. Since the total number of possible faults is very high, fault sampling was adopted for selecting an acceptable number of faults to be injected in the code and data segments and in the processor registers. The number of bit flips injected in each version of the four benchmarks for each fault injection campaign was 30,000. To verify the meaningfulness of the chosen number of faults, several experiments were performed selecting several sets of faults and then comparing the obtained results. Results of each fault injection campaign are shown in Tables *5-1*, *5-2* and *5-3*, which report the average of the results obtained in the different experiments.

Based on the aforementioned procedure, experiments have been performed considering four benchmark programs that are inspired to those in the EEMBC Automotive/industrial suite [151]:

- *5x5 Matrix Multiplication* (MTX): it computes the product of two 5x5 integer matrices.
- *Fifth Order Elliptical Wave Filter* (ELPF): it implements an elliptic filter over a set of 6 samples.
- *Lempel-Ziv-Welch Data Compression Algorithm* (LZW): it compresses data by replacing strings of characters with single codes.
- *Viterbi Algorithm* (V): it implements the Viterbi Algorithm encoding for a 4-byte message.

For each of such benchmarks, up to five different implementations were compared:

- *Plain*: the plain version of the considered benchmark; no hardware or software fault detection techniques are exploited.
- *Software*: the hardened version of the benchmark, obtained using the purely software hardened version combining the approaches described in [2] and [8].
- ED^4I: the hardened version of the benchmark, obtained using the purely software hardening approach described in [152].
- *ABFT:* the hardened version of the MTX benchmark, obtained using the purely software hardening approach described in [153].
- *Hybrid*: the hardened version of the benchmark, obtained using the approach described in this section.

Faults have been classified according to the categories described in Section 1.2.3.

5.2.4.1   Analysis of fault detection capabilities

The following sub-sections report the experimental results gathered with several fault injection campaigns based on the environment described in [150].

5.2.4.2   Faults affecting the code

Results gathered when injecting 30,000 randomly selected single bit flips in the memory area storing the code of each benchmark program (in the 5 considered versions) are reported in Table *5-1*.

When analyzing the reported results about injection into the code segment, the following observations can be made, which relate to the fault classification introduced in Section 4.3. First of all, the reader can easily observe that the software and ED^{4}I approaches are able to significantly reduce the number of faults leading to a wrong answer with respect to the unhardened version: the hybrid approach is always able to further (and significantly) decrease this number. Bit flips affecting the instruction opcode and provoking a wrong answer mainly belong to either the functional_to_functional category (mostly those faults alter the number of requested operands) or branch_to_functional modifications; the hybrid approach shows a higher detection capability with respect to these fault categories than the purely software one, mainly thanks to the check performed at the end of each basic block on the emptiness of the CAM. Such faults may also provoke endless program execution, falling into the timeout case. A detailed analysis of the results summarized in Table I showed that bit flips affecting the operands of an instruction rarely produce a wrong answer effect: both the software and hybrid approaches are able to detect this kind of faults. Additionally, purely software approaches may introduce additional branches to the Program Graph to continuously check the value of the ERROR flag. Moreover, the C compiler may translate some of the C instructions implementing consistency checks as sequences of assembly-level instructions containing new branches. The new branches are not protected with the **test** and **set** functions, and thus some faults may escape software detection techniques. Conversely, when exploiting the hybrid approach, no additional branches are introduced resulting in a lower number of faults leading to wrong answer and time out situations.

The comparison with the ED^{4}I version for Viterbi is not reported in Tables *5-1*, *5-2* and *5-3*. The Viterbi program is mainly based on executing logic operations, but the authors of [152] did not explain how to apply ED^{4}I to such operations (the paper describes how to apply ED^{4}I to arithmetic operations, only).

Regarding ABFT, the results included in Table *5-4* only refers to the MTX program, as it is the only benchmark (among the considered ones) to which this technique can be applied. The coverage obtained by this technique to detect transient faults affecting the code segment is rather low.

5.2.4.3   Faults affecting the data

Table II reports the results gathered when injecting 30,000 randomly selected single bit flips in the memory area storing the data of each benchmark program. These faults are generally very likely not to produce any wrong answer situation when the software approach is adopted; the same happens with the hybrid one. The latter approach performs better than the former when faults producing a time out are considered: this is mainly due to the different behavior with respect to faults belonging to the wrong_branch_condition category.

The ABFT technique fails in detecting some faults affecting the data segment; these escaping faults mainly belong to the wrong_branch_condition category.

5.2.4.4   Faults affecting the processor memory elements

According to the effects they produce, faults in the memory elements within the processor belong either to the wrong_general_purpose_value and wrong_configuration_value categories. The resulting behavior is clearly very different, although both the software and the hybrid approach show low wrong answer figures, as reported in Table *5-3*.

The complete coverage of transient faults affecting the processor memory elements can be reached by using triplication techniques (such as TMR), although this solution is generally undesirable because of the performance reduction, and hardly applicable when the RT-level description of the processor is not available.

5.2.4.5   Overhead analysis

The hybrid approach proposed encompasses three types of overheads with respect to the unhardened version:

- *Area* overhead, related to the adoption of an I-IP.
- *Memory* overhead, due to the insertion in the code of the **I-IPtest()** and **I-IPset()** functions and to the duplication of variables.
- *Performance* overhead, as additional instructions are executed.

In order to quantify the area occupation of the proposed I-IP, the authors of [158] designed it resorting to the VHDL language; the resulting code amountsed to about 450 lines. The I-IP was then synthesized using a commercial tool (Synopsys Design Analyzer) and a generic library. The I-IP was configured to interact with the system bus of the Intel 8051 controller, and it was configured with a CAM memory with 16 entries. The details of the resulting gate-level implementation are shown in Table IV.

When considering the overall hardened system, whose size is the sum of the contributions of the Intel 8051 microcontroller and the related memories, the area overhead introduced by the I-IP is less than 5%. This percent area overhead is expected to further decrease when increasing the complexity of the processor, contrarily to the cost for the triplication of processor memory elements that requires for the analyzed case of study something more than 6% of additional equivalent gates.

To quantify the memory and performance overheads the memory occupation of the programs that were hardened according to the hybrid approach was measured, and then compared with that of the same programs hardened according to the software-based techniques introduced in [154] and [155]. As a reference, the area occupation and program execution time of the original programs was also measured. In Table *5-5* the observed figures are reported. Memory occupation was measured in terms of number of bytes in the data and code segments, while duration was measured in terms of number of clock cycles for program execution.

Results reported in Table *5-5* show that the performance overhead of the hybrid version is about one half in the average than the one of the purely software version.

When considering the memory overhead, we can observe that the increase in the size of the memory required for data is similar in the software and hybrid versions. Conversely, the memory required for the code in the hybrid version is about one half in the average with respect to that required by the software version.

The case of the ELPF program deserves a special attention: this program includes several instructions writing a constant value into a variable. In the software hardened version, this translates into two variables to be written with the same value: the compiler implements this by first loading the value in a register, and then copying the register content into the variables corresponding to the two replicas of the variable. This results in less than duplicating both the code size and the program execution time.

The average block size of the two programs LZW and V is smaller than in the two other programs: this results in a proportionally higher number of I-IPtest() and I-IPset() functions inserted in the code during the hardening phase. For these reason, LZW and V show a higher code overhead figure.

For the same reason, the ratio between branch and functional instructions is higher in LZW and V: since the latter instructions, only, are duplicated in the software and hybrid versions, this results in a relatively low performance overhead for these two programs.

When these figures are coupled with those referring to the area overhead and fault detection capabilities, we can conclude that the hybrid approach is able to effectively improve the dependability of a SoC with limited area overhead, memory increase and performance degradation.

Table 5-1. Fault injection results concerning faults affecting the memory area storing the code.

Prog.	Ver.	Effect-Less		Time-out detected		Software Detected		Failure	
		[#]	[%]	[#]	[%]	[#]	[%]	[#]	[%]
MTX	Plain	20,607	68.6	3,864	12.8	0	0.0	5,529	18.4
	SW	18,798	62.6	2,121	7.0	8,208	27.3	873	2.9
	ABFT	19,356	64.5	3,232	10.7	6,262	20.8	1,150	3.8
	ED^4I	19,356	64.5	3,232	10.7	6,262	20.8	1,150	3.8
	Hybrid	15,567	51.8	3,225	10.7	11,202	37.3	3	0.0
ELPF	Plain	16,071	53.5	3,339	11.1	0	0.0	10,590	35.3
	SW	18,015	60.0	5,583	18.6	5,115	17.0	1,287	4.2
	ED^4I	13,596	45.3	3,283	10.9	13,069	43.5	52	0.1
	Hybrid	14,448	48.1	2,751	9.1	12,750	42.5	51	0.1
LZW	Plain	6,852	22.8	5,469	18.2	0	0.0	17,679	58.9
	SW	10,260	34.2	9,405	31.3	9,420	31.4	915	3.0
	ED^4I	21,890	72.9	2,922	9.7	4,991	16.6	197	0.6
	Hybrid	8,703	29.0	7,836	26.1	13,377	44.5	84	0.2
V	Plain	8,178	27.2	6,093	20.3	0	0.0	15,729	52.4
	SW	10,884	36.2	10,302	34.3	7,518	25.0	1,296	4.3
	ED^4I	N/A							
	Hybrid	10,743	35.8	5,136	17.1	13,734	45.7	387	1.2

Table 5-2. Fault injection results concerning faults affecting the memory area storing the data.

Prog.	Ver.	Effect-Less		Time-out detected		Software Detected		Failure	
		[#]	[%]	[#]	[%]	[#]	[%]	[#]	[%]
MTX	Plain	26,808	89.3	63	0.2	0	0.0	3,129	10.4
	SW	24,930	83.1	57	0.1	5,013	16.7	0	0.0
	ABFT	25,642	85.4	32	0.1	3,818	12.7	508	1.6
	ED^4I	27,029	90.1	48	0.1	2,814	9.3	109	0.3
	Hybrid	25,053	83.5	0	0.0	4,947	16.4	0	0.0
ELPF	Plain	28,623	95.4	33	0.1	0	0.0	1,344	4.4
	SW	22,764	75.8	30	0.1	7,206	24.0	0	0.0
	ED^4I	27,399	91.3	31	0.1	2,503	8.3	67	0.2
	Hybrid	24,339	81.1	0	0.0	5,661	18.8	0	0.0
LZW	Plain	23,889	79.6	450	1.5	0	0.0	5,661	18.8
	SW	18,642	62.1	273	0.9	11,85	36.9	0	0.0
	ED^4I	28,053	93.5	183	0.6	1,482	4.9	282	0.9
	Hybrid	17,958	59.8	0	0.0	12,042	40.1	0	0.0
V	Plain	19,137	63.7	810	2.7	0	0.0	10,053	33.5
	SW	17,067	56.8	450	1.5	12,483	41.9	0	0.0
	ED^4I	Not Available							
	Hybrid	17,433	58.1	0	0.0	12,567	41.9	0	0.0

Table 5-3. Fault injection results concerning faults affecting the memory elements within the procesor.

Prog.	Ver.	Effect-Less		Time-out detected		Software Detected		Failure	
		[#]	[%]	[#]	[%]	[#]	[%]	[#]	[%]
MTX	*Plain*	27,777	92.5	927	3.0	0	0.0	1,296	4.3
	SW	27,987	93.2	189	0.6	1,734	5.7	90	0.3
	ABFT	13,968	93.1	66	0.4	829	5.5	137	0.9
	ED⁴I	13,903	92.6	334	2.2	660	4.4	103	0.9
	Hybrid	27,039	90.1	651	2.1	2,265	7.5	45	0.1
ELPF	*Plain*	27,789	92.6	948	3.1	0	0.0	1,263	4.2
	SW	28,035	93.4	93	0.3	1,641	5.4	231	0.7
	ED⁴I	13,618	90.7	18	0.1	1,328	8.8	36	0.2
	Hybrid	26,817	89.3	807	2.6	2,307	7.6	69	0.2
LZW	*Plain*	26,763	89.2	705	2.3	0	0.0	2,532	8.4
	SW	26,871	89.5	353	1.1	2,623	8.7	153	0.5
	ED⁴I	14,041	93.6	102	0.6	717	4.7	140	0.9
	Hybrid	27,300	91.1	600	2.0	1,920	6.4	90	0.3
V	*Plain*	27,396	91.3	1,437	4.7	0	0.0	1,167	3.8
	SW	27,618	92.0	933	3.1	1,236	4.1	213	0.7
	ED⁴I	Not Available							
	Hybrid	26,907	89.6	939	3.1	2,067	6.8	87	0.2

Table 5-4. I-IP synthesis results summary.

Logic component	Equivalent gates [#]
Bus interface	251
Control Flow Consistency Check	741
Data Consistency Check	1,348
CAM Memory	1,736
TOTAL	**4,076**

Table 5-5. Memory and performance overheads summary.

Prog.	Version	Execution time (CC)		Code size (B)		Data size (D)	
		[#]	[%]	[#]	[%]	[#]	[%]
MTX	*Plain*	13,055	-	329	-	16	-
	Software	42,584	226.1	1,315	299.7	34	112.5
	ABFT	49,792	178.2	768	233.4	32	100.0
	ED⁴I	24,717	189.3	524	59.2	30	87.5
	Hybrid	27,930	113.9	683	107.6	34	112.5
ELPF	*Plain*	12,349	-	384	-	48	-
	Software	46,545	276.9	1,527	297.6	100	108.3
	ED⁴I	23,136	187.3	663	72.6	62	29.1
	Hybrid	21,946	77.7	645	67.9	100	108.3
LZW	*Plain*	19,209	-	232	-	35	-
	Software	92,003	378.9	1,898	718.1	72	105.7
	ED⁴I	35,393	184.2	878	378.4	64	82.8
	Hybrid	38,878	102.3	859	270.2	72	105.7
V	*Plain*	286,364	-	436	-	85	-
	Software			Not Available			
	ED⁴I	598,410	208.97	1,323	203.44	172	102.35

REFERENCES

108. A. Mahmood, E. J. McCluskey, "Concurrent error detection using watchdog processors-a survey", IEEE Transaction on Computers, Vol. 37, No. 2, February 1988, pp. 160-174.

109. F. Rodriguez, J.C. Campelo, J.J. Serrano, "Improving the interleaved signature instruction stream technique", IEEE Canadian Conference on Electrical and Computer Engineering, 2003, Vol. 1, pp. 93 - 96

110. F. Rodriguez, J.C. Campelo, J.J. Serrano, "A Watchdog Processor Architecture with Minimal Performance Overhead", International Conference on Computer Safety, Reliability and Security, 2002, pp. 261-272

111. J. Ohlsson, M. Rimen, "Implicit Signature Checking", Proc. 25th International Symposium on Fault-Tolerant Computing, 1995, pp. 218-227.

112. J. Ohlsson, M. Rimen, U. Gunneflo, "A study of the effects of transient fault injection into a 32-bit RISC with built-in watchdog", Twenty-Second International Symposium on Fault-Tolerant Computing, 1992, FTCS-22, pp. 316 – 325.

113. N.R. Saxena, E.J. McCluskey, "Control Flow Checking Using Watchdog Assists and Extended-Precision Checksums", IEEE Transactions on Computers, Vol. 39, No. 4, Apr. 1990, pp. 554-559.

114. N.R. Saxena, E.J. McCluskey, "Control-flow checking using watchdog assists and extended-precision checksums", Digest of Papers. of Nineteenth International Symposium on Fault-Tolerant Computing, 21-23 June 1989, pp. 428 – 435.

115. N.J. Warter, W.-m.W. Hwu, "A software based approach to achieving optimal performance for signature control flow checking", FTCS-20. Digest of Papers of 20th International Symposium on Fault-Tolerant Computing, 26-28 June 1990 pp. 442 – 449.

116. K. Wilken, J.P. Shen, "Continuous signature monitoring: efficient concurrent-detection of processor control errors", Proc. IEEE International Test Conference, 1988, pp. 914 – 925.

117. K. Wilken, J.P. Shen, "Continuous Signature Monitoring: Low-Cost Concurrent Detection of Processor Control Errors", IEEE Trans. on Computer-Aided Design, Vol. 9, No. 6, June 1990, pp. 629-641.

118. K.D. Wilken, "Optimal signature placement for processor-error detection using signature monitoring", Digest of Papers of Twenty-First International Symposium on Fault-Tolerant Computing, 1991, pp. 326 – 333.

119. K.D. Wilken, "An optimal graph-construction approach to placing program signatures for signature monitoring", IEEE Transactions on Computers, Vol. 42, Issue: 11, Nov. 1993, pp.1372 – 1381.

120. J.P. Shen and M.A. Schuette, "On-line Self-Monitoring Using Signatured Instruction Streams", Proc. IEEE International Test Conference, 1983, 1983, pp. 275-282.

121. M. Namjoo, "Techniques for concurrent testing of VLSI processor operation", in IEEE International Test Conference, 1982, Nov. 15-18, 1982, pp. 461-468.

122. M. Namjoo, "CERBERUS-16: An architecture for a general purpose watchdog processor", Digest of Papers of Thirteenth International Symposium on Fault-Tolerant Computing, 1983, pp. 216-219.

123. M. Namjoo, E.J. McCluskey, "Watchdog processors and capability checking", Digest of Papers of Twelfth International Symposium on Fault-Tolerant Computing, FTCS-12, 1982, pp. 245-248

124. A. Mahmood, D.J. Lu, and E.J. McCluskey, "Concurrent fault detection using a watchdog processor and assertions", Proc. IEEE International Test Conference, 1983, pp. 622-628

125. S.H. Saib, "Distributed architectures for reliability", Proc. AIAA Computer in Aerospace Conference, 1979, pp. 458-462

126. A. Mahmood, A. Ersoz, E.J. McCluskey, "Concurrent system level error detection using a watchdog processor", Proc. IEEE International Test Conference, 1985, pp. 145-152

127. T. Sridhar and S.M. Thatte, "Concurrent checking of program flow in VLSI processors", in IEEE International Test Conference, 1982, Nov. 15-18, 1982, pp. 191-199.

128. A. Benso, S. Di Carlo, G. Di Natale, P. Prinetto, "A watchdog processor to detect data and control flow errors", On-Line Testing Symposium, 2003. IOLTS 2003. 9th IEEE, 7-9 July 2003, pp. 144 – 148

129. J.B. Eifert, J.P. Shen, "Processor Monitoring Using Asynchronous Signatured Instruction Streams", in Dig., 14th Int. Conf. Fault-Tolerant Comput., FTCS-14, Kissimmee, FL, June 20-22, 1984, pp. 394-399

130. S.J. Upadhyaya, B. Ramamurthy, "Concurrent process monitoring with no reference signatures", IEEE Transactions on Computers, Vol.: 43, Issue: 4, April 1994, pp. 475 – 480

131. T. Michel, R. Leveugle and G. Saucier, "A New Approach to Control Flow Checking without Program Modification", Proc. FTCS-21, 1991, pp. 334 – 341.

132. H. Madeira, J. Camoes, J.G. Silva, "A watchdog processor for concurrent error detection in multiple processor system", Microprocessors and Microsystems, Vol. 15, No. 3, April 1991, pp. 123-131

133. X. Delord, G. Saucier, "Control flow checking in pipelined RISC microprocessors: the Motorola MC88100 case study", Proceedings of Euromicro '90 Workshop on Real Time, 6-8 June 1990, pp. 162 – 169.

134. M.Z. Khan, J.G. Tront, "Detection of transient faults in microprocessors by concurrent monitoring", Test Conference, 1989. Proceedings. 'Meeting the Tests of Time'., International, 29-31 Aug. 1989, p. 948

135. S.P. Tomas and J.P. Shen, "A roving monitoring processor for detection of control flow errors in multiple processor systems", in Proc. IEEE Int. Conf. Comput. Design: VLSI Comput., Port Chester, NY, Oct. 7-10, 1985, pp. 531-539.

136. M. Namjoo, "CERBERUS-16: An Architecture for a General Purpose Watchdog Processor", Proc. Symposium on Fault-Tolerant Computing, 1983, pp. 216-219.

137. D.J. Lu, "Watchdog processor and structural integrity checking", IEEE Trans. Computers, vol. C-31, 1982 Jul, pp. 681-685.

138. J. R. Kane and S.S. Yau, "Concurrent software fault detection", IEEE Trans. Software Eng., vol. SE-1, pp. 87-99, Mar. 1975.

139. G. Miremadi, J. Ohlsson, M. Rimen, J. Karlsson, "Use of Time and Address Signatures for Control Flow Checking", International Conference on Dependable Computing for Critical Applications (DCCA-5), 1995, pp. 113-124.

140. H. Madeira and J.G. Silva, "On-line Signature Learning and Checking", Dependable Comp. For Critical Applications, DCCA-2, Springer-Verlag, 1992.

141. B. Ramamurthy, S. Upadhyaya, "Watchdog processor-assisted fast recovery in distributed systems", International Conference on Dependable Computing for Critical Applications (DCCA-5), 1995, pp. 125-134

142. A. Mahmood and E.J. McCluskey, "Watchdog Processor: Error Coverage and Overhead", 15th Ann. Int'l Symp. Fault-Tolerant Computing (FTCS-15), pp. 214-219, June 1985.

143. V.S. Iyengar and L.L. Kinney, "Concurrent fault detection in microprogrammed control units", IEEE Trans. Comput., vol. C-34, pp. 810-821, Sept. 1985.

144. D. J. Lu, "Watchdog processor and VLSI", in Proc. Nat. Electron. Conf., vol. 34, Chicago, IL, Oct. 27-28, 1980, pp. 240-245.

145. S.S. Yau, F.-C. Chen, "An Approach to Concurrent Control Flow Checking", IEEE Transactions on Software Engineering, Vol. SE-6, No. 2, March 1980, pp. 126-137.

146. S.M. Ornstein, W.R. Crowther, M.F. Kraley, R.D. Bressler, A. Michel, and F.E. Heart, "Pluribus – A reliable multiprocessor", in Proc. AFIPS Conf., vol. 44, Anahein, CA, May 19-22, 1975, pp. 551-559.

147. J. R. Connet, E. J. Pasternak, and B.D. Wagner, "Software defenses in real time control systems", in Dig. Int. Symp. Fault Tolerant Comput., FTCS-2, Newton, MA, June 19-21, 1972, pp. 94-99.

148. J. S. Novak and L.S. Tuomenoksa, "Memory mutilation in stored program controlled telephone systems", in Conf. Rec. 1970 Int. Conf. Commun., vol. 2, 1970, pp. 43-32 to 43-45.

149. S.F. Daniels, "A concurrent test technique for standard microprocessors", in Dig. Papers Compcon Spring 83, San Francisco, CA, Feb. 28 – Mar. 3, 1983, pp. 389-394.

150. P. Civera, L. Macchiarulo, M. Rebaudengo, M. Sonza Reorda, M. Violante, "An FPGA-based approach for speeding-up Fault Injection campaigns on safety-critical circuits", Journal of Electronic Testing: Theory and Applications, Vol. 18, No. 3, June 2002, pp. 261-271

151. http://www.eembc.org

152. N. Oh, S. Mitra, E.J. McCluskey, "ED4I: error detection by diverse data and duplicated instructions", IEEE Transactions on Computers, Vol. 51, No. 2 , Feb. 2002, pp. 180-199

153. K. H. Huang, J. A. Abraham, "Algorithm-Based Fault Tolerance for Matrix Operations", IEEE Transaction on Computers, vol. 33, Dec 1984, pp. 518-528

154. P. Cheynet, B. Nicolescu, R. Velazco, M. Rebaudengo, M. Sonza Reorda, M. Violante, "Experimentally evaluating an automatic approach for generating safety-critical software with respect to transient errors", IEEE Transaction on Nuclear Science, Vol. 47, No. 6, December 2000, pp. 2231-2236

155. O. Goloubeva, M. Rebaudengo, M. Sonza Reorda, M. Violante, "Soft-error Detection Using Control Flow Assertions", IEEE Int.l Symp. on Defect and Fault Tolerance in VLSI Systems, 2003, pp. 581-588

156. L. Bolzani, M. Rebaudengo, M. Sonza Reorda, F. Vargas, M. Violante, "Hybrid Soft Error Detection by means of Infrastructure IP cores", IEEE International On-Linc Testing Symposium, 2004, pp. 79-84

157. P. Bernardi, L. Bolzani, M. Rebaudengo, M. Sonza Reorda, F. Vargas, M. Violante, "Hybrid Soft Error Detection by means of Infrastructure IP cores", IEEE International Conference on Dependable Systems and Networks, 2005, pp. 50-58

158. P. Bernardi, L. Bolzani, M. Rebaudengo, M. Sonza Reorda, F. Vargas, M. Violante, "A new Hybrid Fault Detection Technique for Systems-on-a-Chip", accepted for publication on IEEE Transactions on Computer, 2006

159. A. Rajabzadeh, M. Mohandespour, G. Miremadi, "Error Detection Enhancement in COTS Superscalar Processors with Event Monitoring Features", Proc. of the 10-th

IEEE Pacific Rim International Symposium on Dependable Computing, 2004, pp. 49-54

160. A. Rajabzadeh, "Experimental Evaluation of Master/Checker Architecture Using Power Supply- and Software-Based Fault Injection", Proc. of the 10-th IEEE On-Line Testing Symposium, 2004, pp. 239-244

161. A. Rajabzadeh, "A 32-bit COTS-based Fault-Tolerant Embedded System", Proc. Of the 11-th IEEE On-Line Testing Symposium, 2005

Chapter 6

FAULT INJECTION TECHNIQUES

1.   INTRODUCTION

Many approaches have been proposed to perform fault injection, which can be defined as the deliberate insertion of faults into an operational system to observe its response [162]. They can be grouped into simulation-based techniques [163], software-implemented techniques [164][165][166][167], and hybrid techniques, where hardware and software approaches are applied together to optimize the performance [168][169].

Listing and describing all the available approaches is out of the scope of this chapter, whose purpose is to give a synthetic overview of the possible approaches to fault injection. For this reason we decided to present only one approach for each of the aforementioned groups.

Before proceeding with the description of fault-injection techniques (in section 4) we present some background concepts in section 2, and assumptions in section 3.

2.   THE FARM MODEL

In this book we refer to fault injection as a mean to validate dependability measures of a target system constituted by a processor-based hardware architecture and software application.

A good approach to characterize a fault injection environment is to consider the FARM classification proposed in [167]. The FARM attributes are the following:

- *F*: the set of faults to be deliberately introduced into the system.
- *A*: the set of activation trajectories that specify the domain used to functionally exercise the system.
- *R*: the set of readout that corresponds to the behavior of the system.
- *M*: the set of measures that corresponds to the dependability measures obtained trough the fault injection.

The FARM model can be improved by also including the set of workloads *W*.

The measures M can be obtained experimentally from a sequence of fault-injection case studies. An injection campaign is composed of elementary injections, called *experiments*. In a fault-injection campaign the input domain corresponds to a set of faults F and a set of activations A, while the output domain corresponds to a set of readouts R and a set of measures M.

The single experiment is characterized by a fault *f* selected from F and an activation trajectory *a* selected from A in a workload *w* from W. The behavior of the system is observed and constitutes the readout *r*. The experiment is thus characterized by the triple <f, a, r>. The set of measures M is obtained in an injection campaign elaborating the set of readouts R for the workloads in W.

2.1   Fault Injection requirements

The FARM model can be considered as an abstract model that describes the attributes involved in a fault-injection campaign, but it does not consider the fault-injection environment, (i.e., the technique adopted to perform the experiments). The same FARM set can be applied to different fault-injection techniques. Before presenting the techniques described in this chapter, we focus on the parameters that should be considered when setting up a fault-injection environment: intrusiveness, speed, and cost.

2.2   Intrusiveness

The intrusiveness is the difference between the behavior of the original target system and that of the same system when it is the object of a fault-injection campaign. Intrusiveness can be caused by:

- The introduction of instructions or modules for supporting fault injection: as an effect, the sequence of executed modules and instructions is different with respect to that of the target system when the same activation trajectories are applied to its inputs.
- Changes in the electrical and logical setups of the target system, which result in a slow-down of the execution speed of the system, or of some

of its components; this means that during the fault-injection campaign the system shows a different behavior from the temporal point of view; we will call this phenomenon *time intrusiveness*.

- Differences in the memory image of the target system, which is often modified by introducing new code and data for supporting the fault-injection campaign.

It is obvious that a good fault-injection environment should minimize intrusiveness, thus guaranteeing that the computed results can really be extended to the original target system.

2.3   Speed

A fault-injection campaign normally corresponds to the iteration of a high number of fault-injection experiments, each focusing on a single fault and requiring the execution of the target application in the presence of the injected fault. Therefore, the time required by the whole campaign depends on the number of considered faults, and on the time required by every single experiment. In turn, this depends on the time for setting up the experiment, and on the one for executing the application in the presence of the fault.

The speed of the fault-injection campaign can thus be improved by proceeding along one or both of the avenues of attack described in the following sub-sections.

2.3.1   Speeding-up the Single fault-injection experiment

The speed of a fault-injection experiment is computed considering the ratio between the time required by the normal execution (without fault injection) and the average elapsed time required by a single fault-injection experiment. The increase in the elapsed time is due to the operations required to initialize the experiment, to observe the readouts, to inject the fault, and to update the measures.

2.3.2   Reducing the Fault List Size

Since in a given time, the number of possible experiments is limited, a crucial issue when devising a fault-injection environment is the computation of the list of faults to be considered. One challenge is to reduce the large fault space associated with highly integrated systems, improving sampling techniques and models that equivalently represent the effects of low-level faults at higher abstraction levels.

The fault list should be representative enough of the whole set of possible faults that can affect the system, so that the validity of the obtained results is not limited to the faults in the list itself. Unfortunately, increasing the size of the fault list is seldom a viable solution due to the time constraints limiting the maximum duration of the fault-injection experiment. In general, the goal of the fault list generation process is to select a representative sub-set of faults, whose injection can provide a maximum amount of information about the system behavior, while limiting the duration of the fault-injection experiment to acceptable values.

2.4   Cost

A general requirement valid for all the possible target systems is that the cost of the fault-injection environment must be as limited as possible, and negligible with respect to the cost of the system to be validated.

We can consider as a cost the following issues:

- The hardware equipment and the software involved in the fault-injection environment.
- The time required to set up the fault injection environment and to adapt it to the target system.

The first issue is strictly related to the fault injection technique chosen, whereas the second one implies to define a system as flexible as possible that can be easily modified when the target system is changed, and can be easily used by the engineers involved in the fault injection experiments.

3.   ASSUMPTIONS

In this Section we report the assumptions in terms of the FARM model, and choices underlying the organization of the fault-injection environment we will present in the following of this chapter.

3.1   Set F

It is the set of faults to be injected in a fault-injection campaign. First of all, the fault model has to be selected. This choice is traditionally made taking into account from one side the need for a fault model that is as close as possible to real faults, and from the other side the practical usability and manageability of the selected fault model. Based on these constraints, the fault model we selected is the SEU/SET (see Chapter 1 for further details).

Each fault is characterized by the following information:

- *Fault injection time*: it is the time instant when the fault is first inoculated in the system. Depending on the injection methodology, it may be expressed using different unit of measure:
  - Nanoseconds, in the case of simulation-based fault injection.
  - Number of instructions, in case of software-implemented fault injection.
  - Number of clock cycles, in case of hybrid-based fault injection.
- *Fault location*: it is the system's component the fault affects. It may be expressed as the address of the memory location or the register where the SEU has to be injected, or the gate where the SET has to be injected.
- *Fault mask*: in case the faulty component is an n-bit-wide register, the fault mask is the bit mask that selects the bit(s) that has (have) to be affected by the SEU.

A golden-run experiment is performed in advance and is used as a reference for fault-list generation and collapsing. The golden-run can be obtained assuming a deterministic environment, whose behavior can be deterministically determined when the input stimuli are given.

The size of the fault list is a crucial parameter for any kind of fault-injection experiment, because it affects dramatically the feasibility and meaningfulness of the whole fault-injection experiment. For this reason, the presented techniques include a module for fault-list collapsing, which is based on the techniques presented in [170][171]. The rules used to reduce the size of the fault-list do not affect the accuracy of the results gathered through the following fault-injection experiments, but simply aim at avoiding the injection of those faults whose behavior can be foreseen a priori. The validity of the collapsing rules is bounded to the specific fault-injection environment that is going to be used, and to the set of input stimuli the target system is going to receive.

As far as SEUs in processor-based systems are considered, a fault can be removed from the fault list when it can be classified in one of the following classes:

- It affects the operative code of an instruction and changes it into an illegal operative code; therefore, the fault is guaranteed to trigger an error detection mechanism when the instruction is executed (possibly provided by the processor).
- It affects the code of an instruction after the very last time the instruction is executed, and it is thus guaranteed not to generate any effect on the program behavior.
- It affects a memory location containing the program data or a processor register before a write access or after the very last read access; it is thus guaranteed not to generate any effect on the program behavior.

- It corresponds to flipping the same bit of the code of an instruction than another fault, during the period between two executions of that instruction; the two faults thus belong to the same equivalence class, and can thus be collapsed to a single fault.
- It corresponds to flipping the same bit of a memory location containing the program data, or a processor register during the same period between two consecutive accesses of that location than another fault; the two faults thus belong to the same equivalence class, and can thus be collapsed to a single fault.

Experimental results gathered with some benchmark programs show that the average reduction in the fault list size obtained applying the proposed collapsing techniques is about 40% [170], considering an initial fault list composed of a random distribution of faults in the data memory, code memory, and processor registers.

As far as SETs affecting a combinational component, or the combinational part of a sequential component, a fault can be removed from the fault list if its fault-injection time and fault-location are such that its effects cannot reach the circuit outputs in time for being sampled.

Let T_H be the time when the SET is originated by a particle strike, δ be the worst-case SET duration for the considered type of particles, T_S the time when the outputs of the circuit are sampled (determined by the system clock cycle) and $\prod$ is the set of the propagation delays associated to the sensitized paths from the faulty gate to the circuit outputs, e.g., all those paths that, due to the input configuration on the circuit inputs, let a change on the output of the faulty gate to spread the circuit outputs. Any SET is effect-less, i.e., its effects cannot reach the circuit outputs, if the following condition is met:

$$T_H + \delta + t < T_S \qquad \forall\, t \in \prod \qquad (2)$$

If eq. 1 holds, it means that as soon as the SET expires and the expected value is restored on the faulty gate, the correct value has enough time to reach the circuit outputs, and thus the expected output values are sampled. By exploiting this equation, we observed in [171] compaction ratio ranging from 83% up to 95%.

3.2   Set A

Two important issues are related to this point. On the one side it is important to understand how to determine an input trajectory to be applied to the target system during each fault-injection experiment. Several proposals have been made to solve this general problem. In this paper, we do not deal with this problem, but we limit our interest to the techniques for performing the fault-injection campaign, once the trajectory is known. On the other hand, there is the problem of how to practically apply the trajectory to the system. This issue is particularly critical when considering embedded system, since they often own a high number of input signals of different types (digital and analog, high- and low-frequency, etc.).

3.3 Set R

This set of information is obtained by observing the system behavior during each fault injection experiment, and by identifying the differences with respect to the fault-free behavior. Note that all the operations involved by the observation task should also be minimally intrusive.

3.4 Set M

At the end of the fault-injection campaign, a proper tool should build a report concerning the dependability measures and fault coverage computed on the whole fault list. Fault coverage is defined with respect to the possible effects of faults, which were introduced in Chapter 1, and which are report here for the sake of completeness. In this chapter we refer to the following classification.

6. *Effect-less fault.* The fault does not propagate as an error neither as a failure. In this case the fault appeared in the system and remained passive for a certain amount of time, after which it was removed from the system. As an example, let us consider a fault that affects a variable x used by a program. If the first operation the program performs on x after x was affected by the fault is a write operation, then a correct value is overwritten over the faulty one, and thus the system returns in a fault-less state.

7. *Failure.* The fault was able to propagate within the system until it reached the system's output.

8. *Detected fault.* The fault produced an error that was identified and signaled to the system's user. In this case the user is informed that the task the system performs was corrupted by a fault, and the user can thus take the needed countermeasure to restore the correct system functionalities. In systems able to tolerate the presence of faults, the needed countermeasures may be activated automatically. Error detection is performed by means of mechanisms, *error-detection mechanisms*, embedded in the system whose purpose is to monitor the behavior of the system, and to report anomalous situations. When considering a processor-based system, error-detection mechanisms can be found in the processor, or more in general in the hardware

components forming the system, as well as in the software it executes. The former are usually known as *hardware-detection mechanisms*, while the latter are known as *software-detection* mechanisms. As an example of the hardware-detection mechanisms we can consider the *illegal instruction trap* that is normally executed when a processor tries to decode an unknown binary string coming from the code memory. The unknown binary string may be the result of a fault that modified a valid instruction into an invalid one. As an example of the software-detection mechanisms we can consider a code fragment the designers inserted in a program to perform a range check, which is used to validate the data entered by the systems' user, and to report an alert in case the entered data is out of the expected range. To further refine our analysis, it is possible to identify three types of fault detections:

- *Software-detected fault.* A software component identified the presence of an error/failure and signaled it to the user. As an example, we can consider a subprogram that verifies the validity of a result produced by another subprogram stored in a variable x on the basis of range checks. If the value of x is outside the expected range, the controlling subprogram raises an exception.
- *Hardware-detected fault.* A hardware component identified the presence of an error/failure and signaled it to the user. As an example, we can consider a parity checker that equips the memory elements of a processor. In case a fault changed the content of the memory elements, the checker will identify a parity violation and it will raise an exception.
- *Time-out detected faul.* The fault forced the processor-based system to enter in an endless loop, during which the system does not provide any output results. As an example, the occurrence of this fault type may be detected thanks to a watchdog timer that is started at the beginning of the operations of the processor-based system, and that expires before the system could produce any result.

9. *Latent fault.* The fault either remains passive in the system, or it becomes active as an error, but it is not able to reach the system's outputs, and thus it is not able to provoke any failure. As an example, we can consider a fault that modifies a variable x after the last usage of the variable. In this case, x holds a faulty value, but since the program no longer uses x, the fault is unable to become active and propagate through the system.

10. *Corrected fault.* The fault produced an error that the system was able to identify and to correct without the intervention of the user.

4. THE FAULT INJECTION ENVIRONMENTS

This section described three fault-injection environments we developed in the past years. In section 4.1 we describe a simulation-based environment, in 4.2 a software-implemented fault-injection environment, while in section 4.3 a hybrid environment is summarized.

4.1 Simulation-based fault injection

This type of fault injection consists in evaluating the behavior of systems, which are coded in a description language, by means of simulation tools. Fault injection can be implemented in three different ways:

- The simulation tool is enriched with algorithms that allow not only the evaluation of the faulty-free behavior of system, as normally happen in VHDL or Verilog simulators, but also their faulty behaviors. This solution is very popular as far as certain fault models are considered: for example commercial tools exist that support the evaluation of permanent faults like the stuck-at or the delay one [172]. Conversely, there is a limited support of fault models like SEU, or SET, and therefore designers have to rely on prototypical tools either built in-house or provided by universities.

- The model of the analyzed system is enriched with special data types, or with special components, which are in charge of supporting fault injection. This approach is quite popular since it offers a simple solution to implement fault injection that requires limited implementation efforts, and several tools are available adopting it [173][174][175][176]. This solution is popular since it allows implementing fault injection without the need for modifying the simulator used to evaluate the system behavior. Conversely, the model of the system is modified to support fault injection.

- Both the simulation tool and the system model are left unchanged, while fault injection is performed by means of simulation commands. Nowadays, it is quite common to find, within the instruction set of simulators, commands for forcing desired values within the model [177]. By exploiting this feature it is possible to support SEUs and SETs, as well as other fault models.

As an example of a simulation-based fault-injection system we describe the approach presented in [177], whose architecture is depicted in Fig. *6-1*. The main components of this approach are:

- The model of the target system (coded in VHDL language) that describes the functions the system under analysis implements. For the purpose of the described fault-injection system any level of abstraction (system, register transfer, and gate) and any domain of representation (behavioral, or structural) are acceptable. However, the model of the target system should include enough details for allowing meaningful analysis. As an example, in case the user is interested in understanding the effects of SEUs (see Chapter 1), the model of the target system should describe the memory elements of the system.

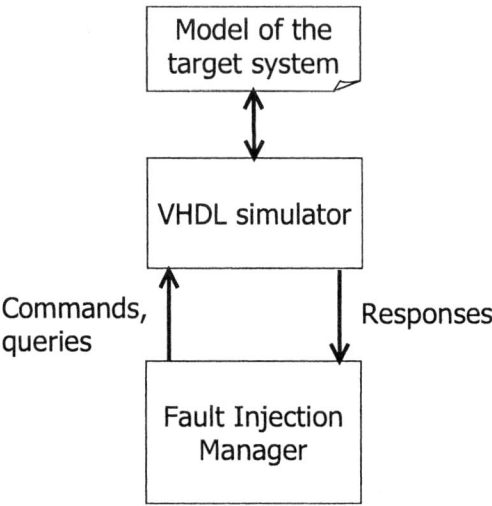

Figure 6-1. An example of simulation-based fault injection

- The VHDL simulator, which is used to evaluate the behavior of the target system. For this purpose any simulation tool supporting the VHDL language, as well as a set of commands allowing monitoring/changing the values of signals and variables during simulation execution is viable.
- The Fault Injection Manager that issues commands to the VHDL simulator to run the analysis of the target system as well as the injection of faults.

Depending of the complexity of the model of the target system, the efficiency of the VHDL simulator adopted, of the workstation used for running the experiments, as well as the number of faults that have to be injected, simulation-based fault-injection experiments may require huge amount of times (many hours if not days) for their execution. In order to overcome this limitation, the approach presented in [177] adopts several techniques aiming at minimizing the time spent for running fault injection.

The approach is composed of three steps:

- *Golden-run execution*: the target system is simulated without injecting any fault and a trace file is produced, gathering information on the system's behavior and on the state of the simulator.
- *Static fault analysis*: given an initial list of faults (fault list) that must be injected, by exploiting the information gathered during golden-run execution we identify those faults whose effects on the system can be determined a-priori, and we remove them from the fault list. Since the injection of each fault encompasses the simulation of the system, by reducing the number of faults that we need to inject we are able to reduce the time needed by the whole experiment.
- *Dynamic fault analysis*: during the injection of each fault, the state of the system under analysis is periodically compared with the golden run at the correspondent time instant. The simulation is stopped as early as the effect of the fault on the system becomes known, e.g., the fault triggered some detection mechanisms, the fault disappeared from the system, or it manifested itself as a failure (see Chapter 1 for a classification of the possible effects of faults). Although the operations needed for comparing the state of the target system with that of the golden run come at a not-negligible cost, the benefits they produce on the time for running the whole experiment are significant. In general, a fault is likely to manifest itself (or to disappear) after few instants since its injection. As a result by monitoring the evolution of the fault for few simulation cycles after its injection, we may be able to stop the simulation execution in advance with respect to the completion of the workload. We can thus save a significant amount of time. Similarly, in case the fault is still latent until few simulation cycles after its injection, it is likely to remain latent, or manifest itself, until the completion of the workload. In this case, the state of the target system and those of the gulden rule are no longer compared, thus saving execution time, until the end of the injection experiment.

In the following section we give more details about the approach introduced in [177].

4.1.1 Golden run execution

The purpose of this step is to gather information on the behavior of the fault-free target system. Given a set of input stimuli (the workload of the system) that will remain constant in the following fault-injection experiments, two sets of information are gathered, one for performing the static fault analysis and one for performing the dynamic fault analysis.

Static fault analysis requires the complete trace of:

- *Data accesses*: whenever a data is accessed, the time, the type of access (read or write) and the address are recorded.
- *Register accesses*: whenever a register is accessed, the time, the register name and the type of access are recorded.
- *Code accesses*: at each instruction fetch, the address of the fetched instruction is stored in a trace file.

We collect the needed information resorting to ad-hoc modules written in VHDL, called code/data watchers, inserted in the system model. This approach is not intrusive, since code/data watchers work in parallel with the system and do not affect its behavior.

Conversely, for performing dynamic fault analysis we periodically stop the simulation and record a snapshot of the system. A snapshot is composed of the content of the processor registers and the data memory at the current simulation time (i.e., the time instant at which the sample is taken).

This approach is effective because allows gathering information on the system with zero intrusiveness. On the other hand, when addressing very large systems, it could require the availability of large amounts of both memory and disk space. As a consequence, the number of snapshots should be carefully selected.

4.1.2   Static fault analysis

Faults are removed from an initial fault list according to two sets of rules, which are applied by analyzing the information gathered during golden run execution.

We remove from the fault list a fault affecting data if it verifies at least one of the following conditions:

- Given a fault f to be injected at time T at address A, we remove f from the fault list if A is never read again after time T; this rule allows removing the faults that do not affect the system behavior.
- Given a fault f to be injected at time T at address A, we remove f from the fault list if the very first operation that involves A after time T is a write operation.

Conversely, we remove a fault affecting the code if it verifies the following condition: given a fault f to be injected at time T at address A, we remove f from the fault list if the address A corresponds to an instruction that is never fetched again after time T. This rule identifies faults that do not produce any effect and whose injection is therefore useless.

4.1.3 Dynamic fault analysis

Dynamic fault analysis is based on the idea of identifying as early as possible the effect of the injected fault during its simulation. As soon as the effect of a fault become evident, we stop the simulation, potentially saving a significant amount of simulation time. The fault-injection procedure we exploit to implement this idea is described in Fig. *6-2*.

The fault-injection procedure starts by setting a set of breakpoints in the VHDL code of the system to capture the following situations:

- *Program completion*: a breakpoint is set so that simulation is stopped after the execution of the last instruction of the program running on the system. This mechanism is useful to early stop the simulation of faults which cause a premature end of the simulated application.
- *Interrupt*: in order to detect asynchronous events, a breakpoint is set to the VHDL statements implementing the interrupt mechanism activation, which is often used to implement hardware and software Error Detection Mechanisms.
- *Time-out*: the simulation is started with a simulation time much higher than the time required for the golden run program completion. A time-out condition is detected if simulation ends and any breakpoints are reached.

After all the required breakpoints have been properly set, we simulate the system up to the injection time, then injection takes place. Injection is done by exploiting the VHDL simulator commands to modify signals/variables in the VHDL source. After injection, the system is simulated up to the time instant corresponding to the first snapshot after injection time. Finally, the system is compared with the golden run, and the following situations are considered:

- *No failure*: the sate of the target system is equal to the golden run; two alternatives are possible:
  1. When injecting in the data area this implies that the fault effects disappeared from the system and that the fault has no effect on the system behavior; the simulation can thus be stopped.
  2. When injecting in the code area, if the faulty instruction is never fetched again we have that the fault effects disappeared from the system and the simulation can be stopped.
- The state of the target system does not match that observed during the golden run; in this case two alternatives are possible:
  1. *Failure*: the fault has affected system outputs (thus causing a failure) and simulation can be stopped.
  2. *Latent fault*: the fault is still present in the system but it did not affect system outputs: further simulations are therefore required.

```
result Inject( SAMPLE *L,  fault F)
{
  set_breakpoints();

  Simulate( F->time );
  FlipBit( F->loc );
  P = get_snapshot( L, F->time );
  do {
    Simulate( P->time );
    res = Compare( P->regs, P->mem);
    if( res == MATCH && F->area == DATA )
      return(NO_FAILURE);
    if( res == MATCH && F->area == CODE )
      if( F->loc is never fetched again )
        return(NO_FAILURE);
    if( res == FAILURE ) return(FAILURE);
    /* res is LATENT */
    P = P->next;
  } while( P != end );
  return(LATENT);
}
```

Figure 6-2. The fault-injection procedure

4.1.4   Checkpoint-based optimizations

The rationale behind this approach is shown in Fig. *6-3*, where the simulated system time is reported above the horizontal axis, while below it we report the CPU time spent to run VHDL simulation.

Given a fault *f* to be injected at time T^S_F, a not-optimized fault-injection tool spends a time equal to T_{setup} to reach injection time. To minimize simulation time, we periodically save the content of simulator data structures in a collection of checkpoint files. A checkpoint file taken at system time T^S stores all the information required to resume the simulation of the system model from time T^S.

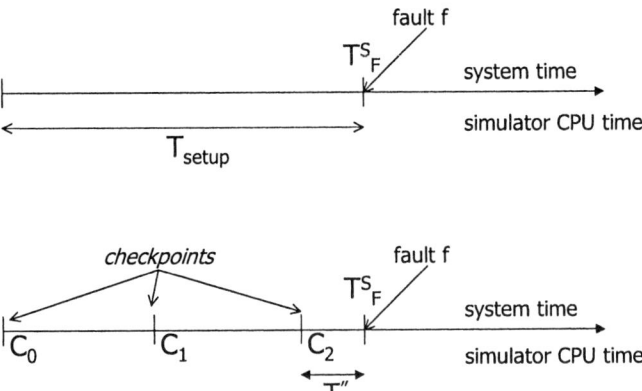

Figure 6-3. Simulator-dependant optimization

When fault f has to be injected, we resume the simulator from the first checkpoint before T^S_F (checkpoint C_2 in the example of Fig. 6-3); therefore, the CPU time spent to reach injection time becomes $T^{''}_{setup}$.

Let T_R be the time for loading the content of a checkpoint file and restoring the simulator data structures, then the following inequality must hold for the approach to be effective:

$$T^{''}_{setup} + T_R < T_{setup} \tag{3}$$

The number of checkpoints should be selected in order to minimize Eq. 3 and to keep the size of checkpoint files below the available disk space.

4.2   Software-implemented fault injection

As an example of a software-implemented fault injection environment we describe the FlexFI system, which was presented in [178], and whose architecture is shown in Fig. 6-4.

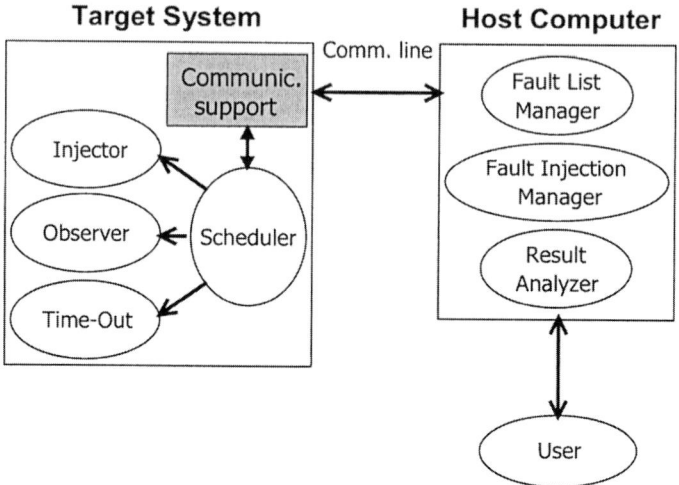

Figure 6-4. The FlexFI fault-injection environment

The system is logically composed of the following main modules:

- The *Fault List Manager* generates the fault list to be injected into the target system.
- The *Fault Injection Manager* injects the faults into the target system;
- The *Result Analyzer* analyzes the results and produces a report concerning the whole Fault Injection campaign.

To minimize the intrusiveness into the target system, the FlexFI system uses a host computer. All the fault-injection tasks which are not strictly required to run on the target system are located on the host computer, which also stores all the data structures (e.g., the Fault List and the output statistics) required by the fault-injection campaign. The host computer communicates with the target system by exploiting the features provided by most systems for debugging purposes (e.g., the serial line handled by a ROM monitor which allows the debugging of most microprocessors).

4.2.1   Fault Injection Manager

The Fault Injection Manager (FIM) is the most crucial part in the whole fault-injection environment. In fact, it is up to the FIM to start the execution of the target application once for each fault of the list generated by the Fault List Manager, to inject the fault at the required time and location, and to observe the system behavior, recovering from any possible failure (e.g., from hardware generated exceptions). The pseudo-code of the FIM is reported in Fig. 6-5.

```
void fault_injection_manager()
{
  campaign_initialization();

    for (every fault fi in the fault list)
  {
    experiment_initialization(fi);

    spawn(target_application);
    spawn(F_I_scheduler);

    wait for experiment completion;

    update_fault_record(fi);
  }
  return();
}
```

Figure 6-5. Fault Injection Manager pseudo-code

During the target application execution, a fault-injection scheduler monitors the advancement of the target program, triggering other fault-injection modules in charge of injecting the fault (Injector module), observing variable values in order to classify the faulty behavior (Observer module), or stop the target application when a time-out condition is reached (Time-out module).

The pseudo-code of the fault-injection scheduler module is reported in Fig. *6-6.* Note that the Observer module refers to an ad hoc data structure, which contains the list of observation points; for each point, this data structure contains the name of the variable, the time when the variable should be observed, as well as the value the variable should have at that time. The list must be filled by the application programmer based on the knowledge of the behavior of the application itself.

```
void F_I_scheduler()
{
  instr_counter++;

  if (instr_counter==fault.time)
      trigger(injector());

  for (i=0; i<num_of_observation_points; i++)
    if (instr_counter==observation_time[i])
      trigger(observer(observed_variable[i], value[i]));

  if (instr_counter>max_time)
      trigger(time_out());
}
```

Figure 6-6. Pseudo-code of the Scheduler module

In order to allow the FIM to maintain the control over the fault-injection campaign, a mechanism has to be devised and implemented to handle the case, in which a hardware exception is activated, and the target application is consequently interrupted. The target system Exception handling procedures have to be suitably modified for this purpose, so that they first communicate to the FIM the type of triggered exception, and then return the control to it (instead of the interrupted instruction).

It is worth underlying the importance of the experiment initialization phase: the effects of the fault injected during an experiment should never affect the behavior of the target application when the following experiment is performed; for this reason, the fault-injection system must restore the environment for the target application execution as a preliminary phase of each experiment. One safe (but slow) way to do so is to restore the full memory image of the application (code and data) and the values of all the relevant system variables. The main issue when implementing this restoring task is to limit its time duration as much as possible, in order to reduce the time requirement of the global fault-injection campaign.

In the following, we will present different techniques for implementing these modules in an embedded system.

4.2.2   Implementation Issues

This solution exploits the trace mode facility existing in most microprocessors for implementing the fault-injection scheduler: thanks to the trace mechanism, a small procedure (corresponding to the fault-injection scheduler) can be activated after the execution of any application assembly instruction with minimum intrusiveness in the system behavior (apart from a slow-down in the application performance). The proposed approach is similar to the ProFI tool [166], with the main difference that the fault-injection experiment is completely executed by the microprocessor without any simulation.

The fault-injection scheduler procedure is in charge of counting the number of executed instructions and verifying whether any fault-injection module reached its activation point. When proper, the procedure activates one of the following modules, each corresponding to a software procedure stored on the target system:

- The Injector module, which is activated when the fault injection time is reached.
- The Time-out module, which is activated when a predefined threshold in terms of number of executed instructions is reached, and stops the target application, returning the control to the FIM located on the host.
- The Observer module, which is in charge of observing the value of target application variables, thus checking whether the application is behaving as in the fault-free fashion or not. When differences are observed, these are communicated to the FIM through the serial interface. The observer module is activated at proper times, depending on the target application characteristics.

We implemented a software-based version of FlexFI for a commercial M68KIDP Motorola board. This board hosts a M68040 microprocessor with a 25Mhz frequency clock, 2 Mbytes of RAM memory, 2 RS-232 Serial I/O Channels, a Parallel Printer Port, and a bus-compatible Ethernet card. To guarantee a deterministic behavior the internal caches have been disabled during the FI campaign.

The Fault Injection Manager is composed of the scheduler procedure, which amounts to about 50 Assembly code lines, of the modified Exception handling routine, which needs about 10 Assembly code lines more than the original one, and of the Initialization procedure, which is written partly in ISO-C and partly in Assembly language and globally amounts to about 200 source lines. Due to the high modularity of the FIM code, the task of adapting it to a new application program can easily be accomplished.

When run on some sample benchmark applications, this version of FlexFI showed a slow-down factor due to Fault Injection of about 25 times.

The software-based version of FlexFI is the most general one (the approach can be implemented on virtually any system) and does not require any special hardware, thus being very inexpensive.

On the other side, this approach has some drawbacks:

- There is some code intrusiveness, due to the need for storing the scheduler procedure, as well as the Injector, Observer, and Time-out procedures, in the target system memory.

- There is also some data intrusiveness, since some small data structures, such as the one for storing the information about the current fault and the observation points must also be stored in the target system memory.
- Forcing the target system to work in Trace mode causes a very high degradation in the execution speed of the application program; thus preventing this approach from being used with real-time embedded systems.

4.3   Hybrid fault injection

As an example of hybrid fault-injection environment we present the FIFA system, which was introduced in [ATS'01], whose flow behind is described in Fig. *6-7*. FIFA is intended for supporting the injection of faults in a processor-based system, which is completely modeled in a hardware-description language (similarly to simulation-based environment). The main novelty of FIFA is to adopt an FPGA-based board to emulate the system, while a computer manages the board operations.

According to the FIFA flow, a software tool is sued to instrument the model of the analyzed system according to the mechanisms described in the following sections. The obtained model is then synthesized and mapped on the FPGA board.

Two hardware platforms are used: a host computer and a FPGA board. The former acts as a master and is in charge of managing Fault Injection campaigns. The latter acts as a slave and is used to emulate the system under analysis. In particular, FIFA exploits a FPGA board where two modules are implemented: the emulated system and the Fault Injection Interface, which allows a host computer to control the behavior of the emulated system.

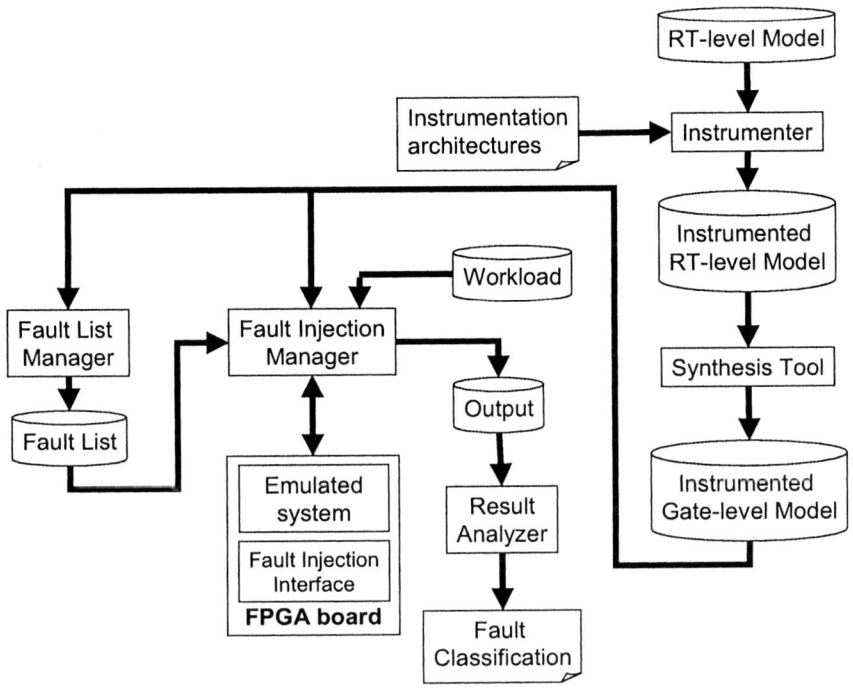

Figure 6-7. The FIFA flow

Three software modules running on the host computer are in charge of performing the typical operations of a Fault Injection environment:

- *Fault List Manager*: it generates the list of faults to be injected in the system.
- *Fault Injection Manager*: it orchestrates the selection of a new fault, its injection in the system, and the analysis of the faulty behavior.
- *Result Analyzer*: it analyzes the behavior of the system during each Fault Injection experiment, categorizes faults according to their effects, and produces statistical information.

4.3.1 The Fault Injection Interface

The Fault Injection Interface executes commands issued by the Fault Injection Manager, running on the host computer, in order to control the behavior of the emulated system.

For the purpose of this paper, the emulated system is a processor core executing a software program. The Fault Injection Interface thus recognizes the following commands:

- *Step*: forces the emulated processor to execute one instruction.
- *Run*: forces the emulated processor to execute a given number of instructions.
- *Evaluate*: sends to the host computer the content of the selected processor storage element.
- *Inject*: modifies the content of a selected processor storage element.
- *Tick*: lets the emulated processor evolve for one clock cycle.

The Step and Run commands implement an instruction-level synchronization strategy, allowing taking control of the emulated processor *after* the execution of an instruction. For example, upon receiving a Step command, the Fault Injection Interface forces the emulated processor to execute one instruction and then waits for further commands from the host computer. Conversely, the Tick command implements a clock-level synchronization strategy, allowing analyzing/modifying the processor behavior *during* the execution of an instruction.

The Evaluate and Inject commands are used to analyze the system state and to perform Fault Injection as described in the following Sub-section.

4.3.2   Injecting Faults

The architecture of a typical processor usually includes the following modules: a processor core comprising the arithmetic/logic and control units embedding both control and internal registers, a general purpose Register File, Instruction and Data caches, and an External Bus used by the processor core to communicate with its peripherals.

In order to perform fault-injection experiments we instrument the processor core model, as shown in Fig. *6-8*, by adding the following modules:

- *Memory Stub logic*: when required, they may isolate the memory from the rest of the system and control its behavior.
- *Bus Stub logic*: as in the previous case, this module is used to take control of the processor External bus, in order to inject faults, apply input stimuli and observe results. In particular, a register M, with the same number of bits of the External Bus, is used to capture the content of the External Bus and send it to the host computer through the Fault Injection Bus. Moreover, it is used to store the masking value for the instrumented bus. At injection time, every bits of M set to logic 1 force the content of the bus to be complemented.

- *Masking logic*: each register in the processor module that is relevant to dependability analysis is connected to an ad-hoc Masking logic. This is in charge of injecting faults and performing fault analysis. Details on the Masking logic can be found in [169].
- *Fault Injection Bus*: it connects all the Masking logic modules inserted in the processor. It includes control signals to access the Masking logic and Stub modules as well as data signals to carry data to and from them. Each Masking logic/Stub module is addressable through the Fault Injection Bus and can be read and written through it.

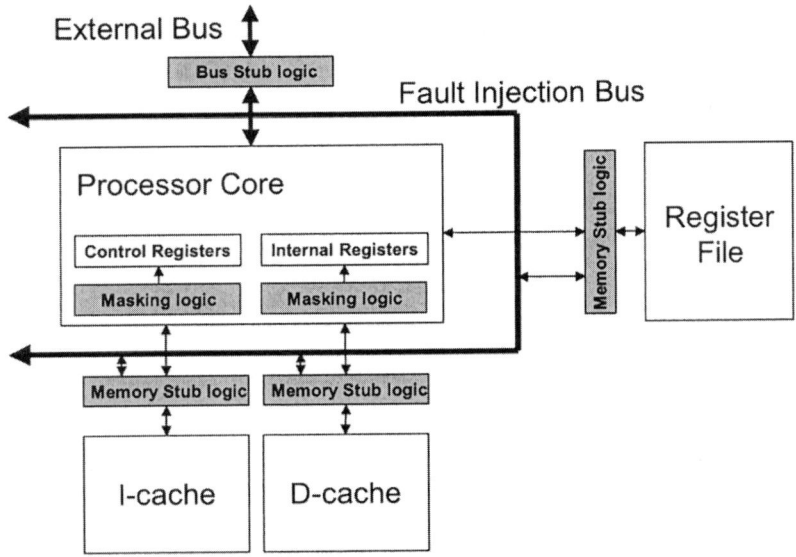

Figure 6-8. A typical processor enriched with Fault Injection features

4.3.3   Memory blocks

Core developers usually adopt a hierarchical approach: a memory module is first described as an isolated entity, resorting either to a behavioral or a structural description, and then it is instantiated wherever needed. Examples of this design style can be found in several cores, such as the PicoJava, and the Intel 8051. A common feature of these memory modules is the presence of address and data buses, as well as the presence of read and write control signals. By driving these signals, we can easily access and possibly alter the content of the memory.

We use a module, called Memory Stub logic, to isolate/control embedded memory modules, according to the architecture reported in Fig. *6-9*. Through the Fault Injection Bus, we are able to take the control of the memory interface, thus we can easily read the memory array content or alter it.

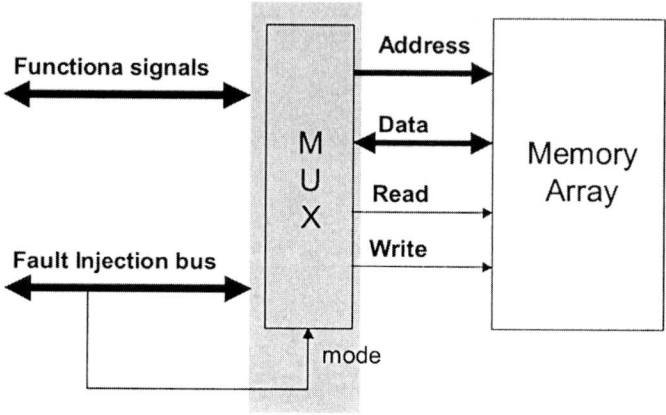

Figure 6-9. Memory Stub logic

Fault injection in memory modules is performed according to the following procedure:
- The Fault Injection Manager leads the emulated system to the injection time by issuing to the Fault Injection Interface the required number of synchronization (Run/Step/Tick) commands.
- The content of the memory location we intend to perturb is read through the Evaluate command and sent to the Fault Injection Manager.
- The Fault Injection Manager computes the faulty value to be injected and writes it back through the Inject command.

4.3.4   Applying stimuli and observing the system behavior

In order to effectively support dependability analysis of safety-critical processor-based systems, the following two classes of applications should be considered:
- *Computing-intensive applications*: they spend most of the execution time for performing computing intensive task, and they commit the results at the end of the computation. As a result, input data should be provided before the activation of the computing algorithm and the amount of

information that should be observed for fault effects classification is mainly dominated by the content of the processor data segment, at the end of the computation.

- *Input/Output-intensive applications*: they spend most of the execution time exchanging data with the environment. Output data are produced during application execution; therefore, they should be continuously recorded in order to classify fault effects. Examples of this family of system are data acquisition systems or communication protocols.

In order to efficiently perform fault-injection experiments, the FPGA board should be equipped with dedicated, high-speed, connection to memory module storing input/output data for each injection experiments. By exploiting this solution, we will boost performances since communication between the FPGA board and the host computer takes place only after a whole Fault Injection campaign (i.e., after several faults have been injected) instead of transmitting information after each fault.

4.3.5 The FI process

The Fault Injection process is composed of the following steps:
1. The circuit description is instrumented according to the previously described transformations.
2. The FPGA board is loaded with the instrumented system description.
3. The Input RAM is programmed with the input data the analyzed system requires.
4. The FPGA-based system is exploited to simulate the fault-free system and the output values at each clock cycle are recorded in the Output RAM: the obtained output trace is the reference trace we use to classify fault effects.
5. For each fault in the fault list, the Fault Injection Manager initializes the FPGA, and performs the injection experiment. The faulty system is lead to injection time, and then a fault is injected by exploiting the procedures described in the previous Sections. Following Fault Injection, the system is emulated up to program completion.
6. At the end of the whole Fault Injection campaign (i.e., after several faults have been injected), the content of Output RAM is sent to the Result Analyzer for fault effects classification.

REFERENCES

162. J. Clark, D. Pradhan, Fault Injection: A method for Validating Computer-System Dependability, IEEE Computer, June 1995, pp. 47-56

163. T.A. Delong, B.W. Johnson, J.A. Profeta III, A Fault Injection Technique for VHDL Behavioral-Level Models, IEEE Design & Test of Computers, Winter 1996, pp. 24-33

164. J. Carreira, H. Madeira, J. Silva, Xception: Software Fault Injection and Monitoring in Processor Functional Units, DCCA-5, Conference on Dependable Computing for Critical Applications, Urbana-Champaign, USA, September 1995, pp. 135-149

165. G.A. Kanawati, N.A. Kanawati, J.A. Abraham, FERRARI: A Flexible Software-Based Fault and Error Injection System, IEEE Trans. on Computers, Vol 44, N. 2, February 1995, pp. 248-260

166. T. Lovric, Processor Fault Simulation with ProFI, European Simulation Symposium ESS95, 1995, pp. 353-357

167. J. Arlat, M. Aguera, L. Amat, Y. Crouzet, J.C. Fabre, J.-C. Laprie, E. Martins, D. Powell, Fault Injection for Dependability Validation: A Methodology and some Applications, IEEE Transactions on Software Engineering, Vol. 16, No. 2, February 1990, pp. 166-182

168. L. T. Young, R. Iyer, K. K. Goswami, A Hybrid Monitor Assisted Fault injection Experiment, Proc. DCCA-3, 1993, pp. 163-174

169. P. Civera, L. Macchiarulo, M. Rebaudengo, M. Sonza Reorda, M. Violante, "Exploiting Circuit Emulation for Fast Hardness Evaluation", IEEE Transactions on Nuclear Science, Vol. 48, No. 6, December 2001, pp. 2210-2216

170. A. Benso, M. Rebaudengo, L. Impagliazzo, P. Marmo, "Fault List Collapsing for Fault Injection Experiments", Annual Reliability and Maintainability Symposium, January 1998, Anaheim, California, USA, pp. 383-388

171. M. Sonza Reorda, M. Violante, "Efficient analysis of single event transients", Journal of Systems Architecture, Elsevier Science, Amsterdam, Netherland, Vol. 50, No. 5, 2004, pp. 239-246

172. TetraMAX, www.synopsys.com

173. E. Jenn, J. Arlat, M. Rimen, J. Ohlsson, J. Karlsson, "Fault Injection into VHDL Models: the MEFISTO Tool", Proc. FTCS-24, 1994, pp. 66-75

174. T.A. Delong, B.W. Johnson, J.A. Profeta III, "A Fault Injection Technique for VHDL Behavioral-Level Models", IEEE Design & Test of Computers, Winter 1996, pp. 24-33

175. D. Gil, R. Martinez, J. V. Busquets, J. C. Baraza, P. J. Gil, "Fault Injection into VHDL Models: Experimental Validation of a Fault Tolerant Microcomputer System", Dependable Computing EDCC-3, September 1999, pp. 191-208

176. J. Boué, P. Pétillon, Y. Crouzet, "MEFISTO-L: A VHDL-Based Fault Injection Tool for the Experimental Assessment of Fault Tolerance", Proc. FTCS'98, 1998

177. B. Parrotta, M. Rebaudengo, M. Sonza Reorda, M. Violante, "New Techniques for Accelerating Fault Injection in VHDL descriptions", IEEE International On-Line Test Workshop, 2000, pp. 61-66

178. A. Benso, M. Rebaudengo, M. Sonza Reorda, "Fault Injection for Embedded Microprocessor-based Systems", Journal of Universal Computer Science (Special Issue on Dependability Evaluation and Validation), Vol. 5, No. 5, pp. 693-711

Index

80386 75

ABFT See Algorithm-Based Fault
   Tolerance
Acceptance test See decider
Active-Stream/Redundant-Stream
   Simultaneous multithreading 56
Algorithm-Based Fault Tolerance 124,
   132, 138, 141
Alpha particle 23, 25, 36
Alternate See Variant
ALU 69
Anti-latchup circuit 40
Assembly-Level Instruction Duplication
   45
Assertions 59, 99
Assigned run-time signature 156, 157
Atomic displacement 24
Availability 9
Available Resource-driven Control flow
   monitoring 79

Babbage, Charles 117
Basic block 63, 144
Best effort recovery 60
Block Entry Exit Checking 76
Block Signature Self Checking 76
Block symbol 93
Branch free interval 63, 86

Caches 49
CAM memory 181
Check interval 83
Checking task operations allocation 80
Checkpass variable 108
Checkpoint 56, 130
Checksum matrix 132
Clock circuitry 49

Code error 13, 16, 17
Code Re-ordering 43
Codeword 32, 33
Commercial-off-the-shelf components 2,
   3
Committed Instructions Counting 175
Comparison vector 119
Computation duplication 37
Conditional instruction 21
Consensus 119
Context switch 56
Control flow checking 63, 85, 95, 155,
   157, 159, 160, 161, 163, 164, 165,
   168, 178
Coronal mass ejection 23
Correction 144

Data checking 180
Data computing block 109
Data Diversity 57
Data error 13, 16, 17
Data integrity 10
Decider 118, 122, 123, 124
Decision point 118
Decoding operation 32, 147, 148
Degradation mechanisms 5
Delay Buffer 56
Derived run-time signature 156, 159
Design diversity 47, 49, 52, 53, 58, 59,
   60, 117, 118
Distributed memory parallel systems 84
Distributed Recovery Block 125, 128
Diversity factor 57
Domains 43
DRB See Distributed Recovery Block
Duplication 48, 142, 147, 174
Dynamic error 11, 19

Elastic collision 25